ESSENTIAL
MATHEMATICAL
SKILLS

A UNSW Press book

Published by
University of New South Wales Press Ltd
University of New South Wales
UNSW Sydney NSW 2052
AUSTRALIA
www.unswpress.com.au

First published 2002

National Library of Australia
Cataloguing-in-Publication entry:

Barry, Steven Ian.
Essential mathematical skills for engineering, science and applied mathematics.

Includes index.
ISBN 0 86840 565 5.

1. Mathematics. 2. Mathematics — Problems, exercises, etc. I. Davis, Stephen. II. Title.

510

Printer BPA

ESSENTIAL MATHEMATICAL SKILLS

for engineering, science and applied mathematics

Dr Steven Ian Barry
+
Dr Stephen Alan Davis

UNSW PRESS

CONTENTS

PREFACE

TO THE STUDENT

There are certain mathematical skills that are essential for any of your courses that use mathematics. Your lecturer will assume that you know them perfectly — not just a vague idea, but that you have completely mastered these skills. Without these necessary skills, you will find present and later subjects extremely difficult. You may also lose too many marks making 'silly' mistakes in exams.

So what skills do you need to have?

This book contains the mathematical skills we think are essential for you to not only know *but remember*. It is not a textbook and does not attempt to teach you, hence there are no long wordy explanations. This book should act as a reminder to you of material you have already learned. If you are having trouble with a section or chapter then we suggest you consult a more thorough textbook. We have left a number of blank pages at the back of the book for you to add in skills that you or your lecturers think are important to remember but we did not include.

This book covers the essential mathematics in the first one to two years of a science, engineering or applied mathematics degree. If you are in a first year undergraduate course you may not have covered some of the material included in this book.

As a guide, we expect our students at University College to have mastered (by the *start* of each semester) the following:

- **First Year — Semester One**: Chapters 1–3.
- **First Year — Semester Two**: Chapters 1–7.
- **Second Year**: Chapters 1–10.
- **Third Year**: Everything in the book!

There are practice tests in Chapter 13 based on these divisions.

Can you do the practice test at the end of these notes?

If you can't then perhaps there are some skills you need to do some revision on. If you can then you may need this book to help you revise those skills later on.

If you want more questions to practice on then see our extensive website:

http://www.ma.adfa.edu.au/~sib/EMS.html

It contains extra questions, fully worked solutions, practice tests and also code for the Maple algebraic manipulation package giving solutions for every example and question.

TO THE LECTURER

What do you assume your students know? What material do you expect them to have a vague idea about (say the proof of Taylor's Theorem) and what material do you want students to know thoroughly (say the derivative of $\sin x$)? This book is an attempt to define what material students should have completely mastered at each year in an applied mathematics, engineering or science degree. Naturally we would like our students to know more than the bare essentials detailed in this book. However, most students do not get full marks in their previous courses and a few weeks after the exam will only remember a small fraction of a course. They are also doing many other courses not involving mathematics and are not constantly using their mathematical skills. This book can then act as guide to what material should realistically be remembered from previous courses. Naturally both the material and the year in which the students see this material will vary from university to university. This book represents what we feel is appropriate to our students during their degrees.

We invite you to look at our extensive web site:

http://www.ma.adfa.edu.au/~sib/EMS.html

It contains more questions, solutions, practice tests and Maple code. There is a database of questions in LaTeX and pdf, which you can use to format your own tests and assignments. We are not concerned that students may access this database; if they can do the questions in the database then they have, in effect, learned the necessary skills.

If you have any questions or queries please do not hesitate to email us.

Steven Barry and Stephen Davis
School of Mathematics and Statistics
University College, UNSW
Canberra, ACT, 2600
email: s.barry@adfa.edu.au

CHAPTER 1

ALGEBRA AND GEOMETRY

1.1 ELEMENTARY NOTATION

1. $\{\}$: A set of objects.
2. $\in$: A member of a set. For example $3 \in \{1, 2, 3\}$.
3. R: The set of real numbers. For example $-1, 3, 3.2, \sqrt{2} \in R$.
4. Z: The set of integers. For example $-2, 0, 3 \in Z$.
5. $<, >$: Less than, greater than. For example $5 < 6$, $7 > 5$.
6. $\leq, \geq$: Less than or equal to, greater than or equal to.
7. $\Longrightarrow$: Becomes. For example $x - 2 = 3 \implies x = 5$.
8. $[a, b]$: Bounds of a variable. For example $x \in [1, 3]$ means $1 \leq x \leq 3$.
9. (a, b): Bounds of a variable. For example $x \in (1, 3)$ means $1 < x < 3$.
10. $\rightarrow$: Tends to. For example $1/x \rightarrow 0$ as $x \rightarrow \infty$.
11. $\approx$: Approximately equal to. For example $3.02 \approx 3$.

EXAMPLES

1. $W = \{f(x) = a + bx : a, b \in R\}$ means W is the set of all functions $f(x) = a + bx$ where a, b are real numbers (constants). Hence $1 + 2x \in W$ and $3 - 1.2x \in W$.
2. $S = \{x : x \geq 5, x \in R\}$ means that S is the set of all numbers bigger than or equal to 5. This is also written as $x \in [5, \infty)$.

1.2 FRACTIONS

A fraction is of the form $\frac{a}{b}$ where a is called the *numerator* and b is called the *denominator*.

Rules for operating on fractions

1. $\frac{a}{c} + \frac{b}{c} = \frac{a+b}{c}$ $\quad (c \neq 0)$

2. $\frac{a}{b} + \frac{c}{d} = \frac{ad+bc}{bd}$ $\quad (b, d \neq 0)$

3. $\frac{a}{c} \times \frac{b}{d} = \frac{ab}{cd}$ $\quad (c, d \neq 0)$

4. $\frac{a}{b} \div \frac{c}{d} = \frac{a}{b} \times \frac{d}{c} = \frac{ad}{bc}$ $\quad (b, c, d \neq 0)$

EXAMPLES

1. $\frac{2}{9} \times \frac{3}{8} = \frac{6}{72} = \frac{1}{12}$

2. $\frac{1}{3} + \frac{1}{6} = \frac{2}{6} + \frac{1}{6} = \frac{3}{6} = \frac{1}{2}$

3. $\frac{x+2}{x-2} - \frac{x-2}{x+2} = \frac{(x+2)^2 - (x-2)^2}{(x-2)(x+2)} = \frac{(x^2+4x+4) - (x^2-4x+4)}{(x^2-4)} = \frac{8x}{x^2-4}$

4. To rearrange the equation $\frac{1}{x} + \frac{1}{y} = \frac{1}{10}$ to find y write

$$\frac{1}{y} = \frac{1}{10} - \frac{1}{x}$$

$$\Longrightarrow \quad \frac{1}{y} = \frac{x-10}{10x} \qquad \textbf{NOT} \quad y = 10 - x$$

$$\Longrightarrow \quad y = \frac{10x}{x-10}.$$

1.3 MODULUS

The absolute value or modulus of x, written $|x|$, is defined by

$$|x| = \begin{cases} x, & \text{if } x \geq 0 \\ -x, & \text{if } x < 0. \end{cases}$$

The absolute value is the magnitude of a number and ignores whether it is positive or negative.

EXAMPLES

1. $|+5| = 5$
2. $|-3| = 3$
3. $|-x|\,|y| = |x|\,|y| = |xy|$

1.4 INEQUALITIES

1. If $x > y$ then $x + a > y + a$ for any a.
2. If $x > y$ then $ax > ay$ if a is positive, but $ax < ay$ if a is negative.
3. If $x > y$ and $u > v$, then $x + u > y + v$.

EXAMPLES

1. To find x such that $-5x - 2 \leq 3$ write

$$\begin{aligned} & -5x - 2 \leq 3 \\ \Longrightarrow \quad & -5x \leq 5 \\ \Longrightarrow \quad & x \geq -1. \end{aligned}$$

2. To find values of x such that $x + 1 > 2x - 5$ we write

$$\begin{aligned} & x + 1 > 2x - 5 \\ \Longrightarrow \quad & x < 6. \end{aligned}$$

Inequalities with modulus

1. The inequality $|x-b|<a$ can be written as $-a<x-b<a$.
2. The inequality $|x-b|>a$ can be written as $x-b>a$ **or** $(x-b)<-a$.

EXAMPLES

1. To find x such that $|2x-1|\leq 3$ write

$$\begin{aligned}|2x-1|\leq 3 &\implies -3\leq 2x-1\leq 3\\ &\implies -2\leq 2x\leq 4\\ &\implies -1\leq x\leq 2.\end{aligned}$$

2. To find x such that $\left|\dfrac{3x-1}{4}\right|\geq 2$ write

$$\begin{aligned}\left|\frac{3x-1}{4}\right|\geq 2 &\implies \frac{3x-1}{4}\geq 2 \quad\text{or}\quad \frac{3x-1}{4}\leq -2\\ &\implies 3x\geq 9 \quad\text{or}\quad 3x\leq -7\\ &\implies x\geq 3 \quad\text{or}\quad x\leq -\frac{7}{3}.\end{aligned}$$

1.5 EXPANSION AND FACTORISATION

$$\begin{aligned}(a+b)(c+d) &= a(c+d)+b(c+d)=ac+ad+bc+bd\\ (a-b)(a+b) &= a^2-b^2\\ (a\pm b)^2 &= a^2\pm 2ab+b^2\end{aligned}$$

EXAMPLES

1. $(x^2-3)^2=x^4+2(-3)x^2+9=x^4-6x^2+9$

2.
$$\begin{aligned}(x-3)(x+5)^2(x+3) &= (x-3)(x+3)(x+5)^2\\ &= (x^2-9)(x^2+10x+25)\\ &= x^4+10x^3+16x^2-90x-225\end{aligned}$$

3.
$$\begin{aligned}\frac{s^2-4}{2+s} &= \frac{(s-2)(s+2)}{2+s}\\ &= s-2\end{aligned}$$

4.
$$\begin{aligned}(a+1)^3 &= (a+1)(a^2+2a+1)\\ &= a^3+3a^2+3a+1\end{aligned}$$

1.5.1 BINOMIAL EXPANSION

$$(a+b)^n = a^n + na^{n-1}b + \frac{n(n-1)}{2!}a^{n-2}b^2 + \cdots + nab^{n-1} + b^n$$

(See also Section 1.13). To remember the coefficients of each term use Pascal's triangle where each number is the sum of the two numbers above it.

$$\begin{array}{ccccccccccc} &&&&&1&&&&& \\ &&&&1&&1&&&& \\ &&&1&&2&&1&&& \\ &&1&&3&&3&&1&& \\ &1&&4&&6&&4&&1& \\ 1&&5&&10&&10&&5&&1 \end{array}$$

Each term in a row represents the coefficients of the corresponding term in the expansion.

EXAMPLES

1. $(a+b)^3 = a^3 + 3a^2b + 3ab^2 + b^3$
2. $(1+x)^4 = 1 + 4x + 6x^2 + 4x^3 + x^4$
3. The coefficient of x^3 in $(2+x)^5$ is $10 \times 2^2 = 40$.

1.5.2 FACTORISING POLYNOMIALS

Factorising a polynomial is the opposite of the expansion described above, that is, splitting the polynomial into its factors:

$$p(x) = (x - a_1)(x - a_2) \cdots (x - a_n).$$

EXAMPLES

1. $x^2 - 1 = (x - 1)(x + 1)$
2. $x^2 - 3x + 2 = (x - 2)(x - 1)$
3. $3x^2 - 7x + 2 = (3x - 1)(x - 2)$
4. $x^3 - 4x^2 + 4x = x(x - 2)^2$
5. $a^3 + 3a^2 + 3a + 1 = (a + 1)^3$

1.6 PARTIAL FRACTIONS

It is sometimes convenient to write

$$\frac{cx + d}{(x + a)(x + b)} = \frac{A}{x + a} + \frac{B}{x + b}$$

where A and B are constants found by equating the numerators of both sides once the right hand side is written as one fraction:

$$cx + d = A(x + b) + B(x + a).$$

Some similar partial fraction expansions are

$$\begin{aligned} \frac{1}{(x + a)^2(x + b)} &= \frac{A}{x + a} + \frac{B}{(x + a)^2} + \frac{C}{x + b} \\ \frac{1}{(x^2 + bx + c)(x + a)} &= \frac{Ax + B}{x^2 + bx + c} + \frac{C}{x + a}. \end{aligned}$$

EXAMPLES

1. Writing $\frac{1}{(x+1)(x-1)}$ in the form $\frac{A}{x+1}+\frac{B}{x-1}$ implies

$$A(x-1)+B(x+1)=1.$$

The constants A and B can be found two simple ways. First, setting

$$\begin{aligned} x=1 &\implies B=\frac{1}{2} \\ x=-1 &\implies A=-\frac{1}{2}. \end{aligned}$$

Alternatively the equation could be expanded as

$$Ax+Bx-A+B=1$$

and the coefficients of x^1 and x^0 equated giving

$$\begin{aligned} A+B&=0 \\ -A+B&=1. \end{aligned}$$

Solving these equations simultaneously gives $A=-1/2$ and $B=1/2$. Thus

$$\frac{1}{(x+1)(x-1)}=\frac{1}{2}\left(\frac{1}{x-1}-\frac{1}{x+1}\right).$$

2. To expand $\frac{3x+1}{(x+7)(x-3)}$ using partial fractions write

$$\frac{3x+1}{(x+7)(x-3)}=\frac{A}{x+7}+\frac{B}{x-3}$$

giving

$$A(x-3)+B(x+7)=3x+1.$$

Setting $x=3$ implies $B=1$ and setting $x=-7$ implies $A=2$. Alternatively, equating the coefficients of

$$Ax+Bx-3A+7B=3x+1$$

gives

$$\begin{aligned} A+B&=3 \\ -3A+7B&=1. \end{aligned}$$

These simultaneous equations are solved for A and B to give $A=2$ and $B=1$. Hence

$$\frac{3x+1}{(x+7)(x-3)}=\frac{2}{x+7}+\frac{1}{x-3}.$$

3. The partial fraction for $\dfrac{1}{(x+1)^2(x+2)}$ is

$$\frac{1}{(x+1)^2(x+2)} = \frac{A}{x+1} + \frac{B}{(x+1)^2} + \frac{C}{x+2}$$

giving

$$1 = A(x+1)(x+2) + B(x+2) + C(x+1)^2$$

so that

$$\begin{aligned} x = -1 \quad &\Longrightarrow \quad 1 = B \\ x = -2 \quad &\Longrightarrow \quad 1 = C \\ \text{order } x^2 \quad &\Longrightarrow \quad 0 = A + C \quad \Longrightarrow \quad A = -1. \end{aligned}$$

Thus

$$\frac{1}{(x+1)^2(x+2)} = \frac{1}{(x+1)^2} - \frac{1}{x+1} + \frac{1}{x+2}.$$

4. The partial fraction for $\dfrac{3}{(x^2+x+1)(x+2)}$ is

$$\frac{3}{(x^2+x+1)(x+2)} = \frac{Ax+B}{x^2+x+1} + \frac{C}{x+2}$$

giving

$$3 = (Ax+B)(x+2) + C(x^2+x+1).$$

Hence

$$\begin{aligned} x = -2 \quad &\Longrightarrow \quad C{=}1 \\ x = 0 \quad &\Longrightarrow \quad 3{=}2B + C \quad \Longrightarrow \quad B = 1 \\ \text{order } x^2 \quad &\Longrightarrow \quad 0{=}A + C \quad \Longrightarrow \quad A = -1. \end{aligned}$$

Thus

$$\frac{3}{(x^2+x+1)(x+2)} = \frac{1}{x+2} - \frac{x-1}{x^2+x+1}.$$

1.7 POLYNOMIAL DIVISION

Polynomial division is a type of long division for polynomials best illustrated by the following examples.

EXAMPLES

1. When dividing $x^2 + 3x + 4$ by $x + 1$ consider only the leading order terms to begin with. Thus x goes into x^2, x times. Thus $x(x + 1) = x^2 + x$, which is subtracted from $x^2 + 3x + 4$. The first step is therefore

$$\begin{array}{r} x \\ x+1\,\overline{)\,x^2+3x+4} \\ \underline{x^2+x} \\ 2x+4 \end{array}$$

The division is completed by considering that x (the leading order of $x + 1$) goes into $2x + 4$ two times. Subtracting $2(x + 1)$ from $2x + 4$ gives

$$\begin{array}{r} x+2 \\ x+1\,\overline{)\,x^2+3x+4} \\ \underline{x^2+x} \\ 2x+4 \\ \underline{2x+2} \\ 2 \end{array}$$

Thus $$\frac{x^2 + 3x + 4}{x + 1} = (x + 2) + \frac{2}{x + 1}.$$

2. Dividing $3x^3 + 2x^2 + x + 1$ by $x - 1$ gives

$$\frac{3x^3 + 2x^2 + x + 1}{x - 1} = 3x^2 + 5x + 6 + \frac{7}{x - 1}.$$

3. $$\frac{4x^3 + 6x^2 + 4x + 1}{2x + 1} = 2x^2 + 2x + 1$$

1.8 SURDS

A *surd* is of the form $\sqrt[n]{a}\ (= a^{1/n})$:

1. $\sqrt{a} \times \sqrt{b} = \sqrt{ab}$
2. $\dfrac{\sqrt{a}}{\sqrt{b}} = \sqrt{\dfrac{a}{b}}$
3. $b\sqrt{a} \pm c\sqrt{a} = (b \pm c)\sqrt{a}$

EXAMPLES

1. $\sqrt{5} \times \sqrt{2} = \sqrt{10}$
2. $\sqrt{27} = \sqrt{9 \times 3} = 3\sqrt{3}$
3. $3\sqrt{10} - 2\sqrt{10} = \sqrt{10}$
4. $\dfrac{\sqrt{14}}{\sqrt{2}} = \sqrt{\dfrac{14}{2}} = \sqrt{7}$

1.8.1 RATIONALISING SURD DENOMINATORS

For an expression of the form

$$\frac{a}{b+\sqrt{c}}$$

it may be preferable to have a rational denominator. A surd denominator is *rationalised* by multiplying the expression by $\dfrac{b-\sqrt{c}}{b-\sqrt{c}}$ (= 1):

$$\begin{aligned}\frac{a}{b+\sqrt{c}} &= \frac{a}{b+\sqrt{c}} \times \frac{b-\sqrt{c}}{b-\sqrt{c}} \\ &= \frac{a(b-\sqrt{c})}{b^2-c}.\end{aligned}$$

EXAMPLES

1.
$$\frac{5}{1+\sqrt{5}} = \frac{5}{1+\sqrt{5}} \times \frac{1-\sqrt{5}}{1-\sqrt{5}}$$
$$= \frac{5-5\sqrt{5}}{(1)^2-(\sqrt{5})^2}$$
$$= \frac{5-5\sqrt{5}}{(-4)}$$
$$= \frac{5\sqrt{5}-5}{4}$$

2.
$$\frac{6x}{1+2\sqrt{x}} = \frac{6x}{1+2\sqrt{x}} \times \frac{1-2\sqrt{x}}{1-2\sqrt{x}}$$
$$= \frac{6x-12x\sqrt{x}}{1-4x}$$

1.9 QUADRATIC EQUATION

A quadratic equation is of the form

$$y = ax^2 + bx + c$$

where a, b, c are constants. The roots of a quadratic equation (when $y = 0$) are

$$x_1, x_2 = \frac{-b \pm \sqrt{b^2-4ac}}{2a}.$$

A quadratic is factorised if it is written in the form

$$y = a(x-x_1)(x-x_2).$$

EXAMPLES

1. The solutions to $x^2 + 3x + 1 = 0$ are

$$x = \frac{-3+\sqrt{5}}{2} \quad \text{or} \quad \frac{-3-\sqrt{5}}{2}.$$

2. The quadratic $y = x^2 + x - 6$ is factorised into $y = (x+3)(x-2)$.
3. The quadratic $y = x^2 + 2x + 1$ is factorised into $y = (x+1)^2$.

4. The solutions to $3x^2 + 5x + 1 = 0$ are

$$x = \frac{-5 \pm \sqrt{25 - 12}}{6}$$

so that

$$x = \frac{-5 + \sqrt{13}}{6} \quad \text{or} \quad \frac{-5 - \sqrt{13}}{6}.$$

1.10 SUMMATION

The summation sign $\sum$ is defined as

$$\sum_{i=1}^{n} f(i) = f(1) + f(2) + f(3) + \cdots + f(n-1) + f(n).$$

EXAMPLE $\sum_{i=1}^{4} i^2 = 1^2 + 2^2 + 3^2 + 4^2 = 30$

1.11 FACTORIAL NOTATION

The factorial notation is defined as follows:

$$n! = n.(n-1).(n-2)\ldots 3.2.1 \qquad \text{where } n \text{ is an integer.}$$

EXAMPLES

1. $5! = 5 \times 4 \times 3 \times 2 \times 1 = 120$
2. $0! = 1$ by definition.
3. $n! = n(n-1)!$
4. $2.4.6.8\ldots 2n = (2.2.2\ldots 2)(1.2.3\ldots n) = 2^n n!$

1.12 PERMUTATIONS

A **permutation** is a particular ordering of a set of unique objects. The **number of permutations** of r unique objects, chosen from a group of n, is given by

$$P_r^n = \frac{n!}{(n-r)!}.$$

EXAMPLE

The number of ways a batting lineup of 3 can be chosen from a squad of 8 cricket players is given by

$$P_3^8 = \frac{8!}{(8-3)!} = \frac{8!}{5!} = 8 \times 7 \times 6 = 336.$$

1.13 COMBINATIONS

If order is not important when choosing r things from a group of n then the number of possible **combinations** is given by

$$C_r^n = \frac{n!}{r!(n-r)!}.$$

EXAMPLES

1. The number of possible groups of 4 delegates chosen from a group of 11 is given by

$$C_4^{11} = \frac{11!}{4!(11-4)!} = \frac{11!}{4!7!} = \frac{11 \times 10 \times 9 \times 8}{4 \times 3 \times 2 \times 1} = 330.$$

2. The number ways of choosing a team of 5 people from 7 is $C_5^7 = 21$.

1.14 GEOMETRY

The trigonometric ratios can be expressed in terms of the sides of a right-angled triangle:

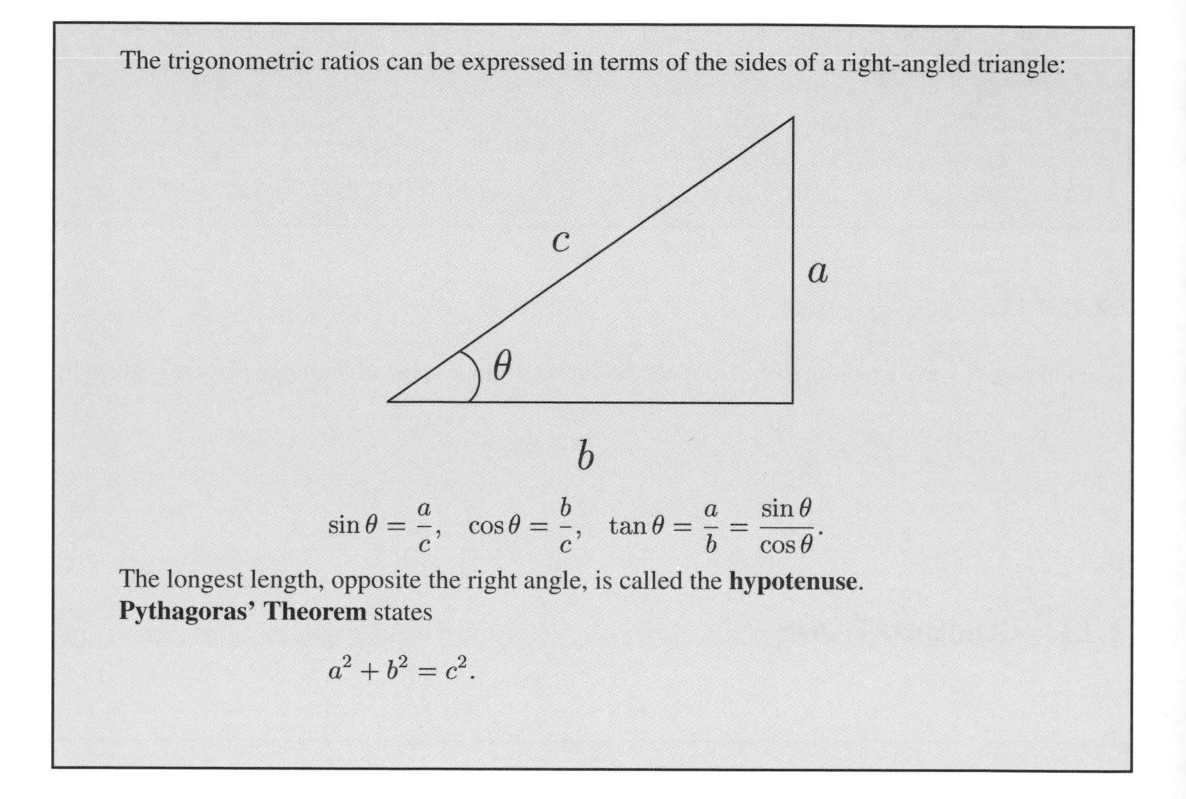

$$\sin\theta = \frac{a}{c}, \quad \cos\theta = \frac{b}{c}, \quad \tan\theta = \frac{a}{b} = \frac{\sin\theta}{\cos\theta}.$$

The longest length, opposite the right angle, is called the **hypotenuse**.
Pythagoras' Theorem states

$$a^2 + b^2 = c^2.$$

The sine, cosine and tangent of the common angles can be related to the following triangles:

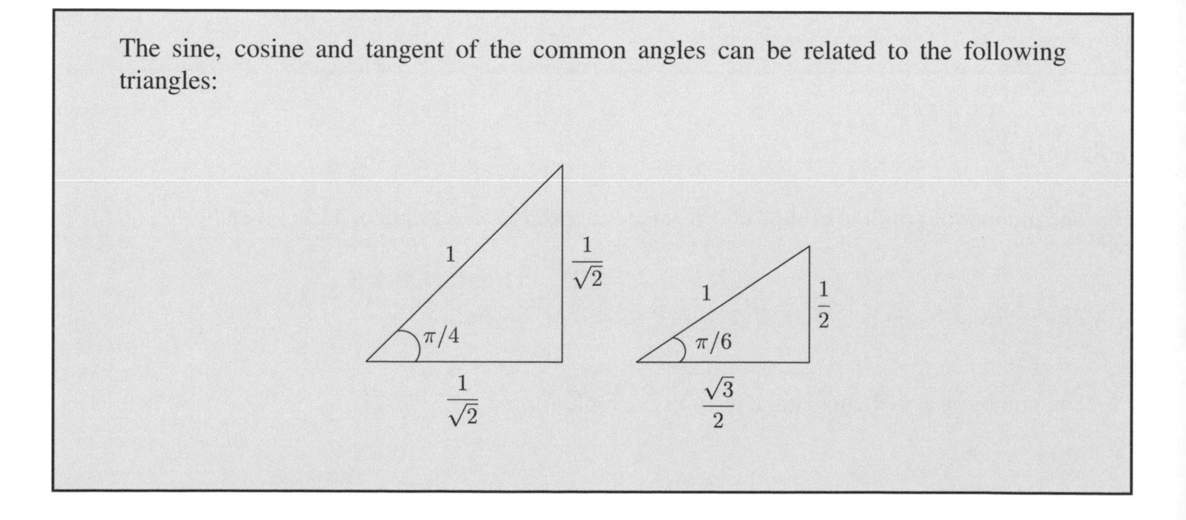

EXAMPLE

$$\sin\frac{\pi}{4} = \frac{1}{\sqrt{2}}, \quad \sin\frac{\pi}{3} = \frac{\sqrt{3}}{2}, \quad \sin\frac{\pi}{6} = \frac{1}{2}, \quad \cos\frac{\pi}{6} = \frac{\sqrt{3}}{2}.$$

The three common triangles are the

1. **isosceles**: any two sides are of equal length.
2. **equilateral**: all three sides of of equal length.
3. **right angled**: one of the angles is $\frac{\pi}{2}$.

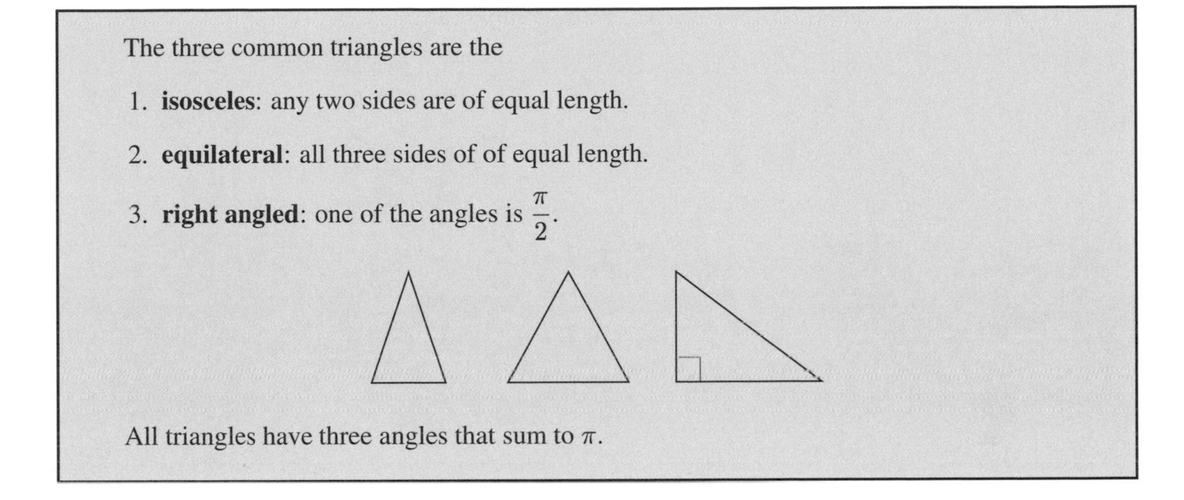

All triangles have three angles that sum to π.

EXAMPLES

1. A right angled triangle has one other angle $\frac{\pi}{6}$. Hence the third angle is $\frac{\pi}{3}$.
2. An equilateral triangle must have three identical angles of $\frac{\pi}{3}$.

1.14.1 CIRCLES

A circle of radius r has

1. **area** $= \pi r^2$
2. **circumference** $= 2\pi r$

EXAMPLES

1. The area of a circle with diameter $d = 6$ is $\pi 3^2 = 9\pi$.
2. The circumference of the circle with diameter $d = 7$ is 7π.

1.15 EXAMPLE QUESTIONS

(Answers are given in Chapter 14)

1. Simplify the following.

   (i) $\frac{1}{2} - \frac{5}{6} + \frac{1}{10}$

   (ii) $\frac{x}{x-3} - \frac{5}{x+2}$

   (iii) $\frac{x-1}{x+2} - \frac{2x}{x-2}$

   (iv) $\frac{2x+1}{x-4} - \frac{x-1}{x+2}$

   (v) $\frac{x^2-2}{x-1} + \frac{x+3}{x}$

   (vi) $\frac{x^3-x^2}{x(x^2-1)}$

2. Find the solution set for the following inequalities.

   (i) $2d + 2 \leq 4d - 3$

   (ii) $3d - 2 > 4d + 6$

   (iii) $|x - 10| < 5$

   (iv) $|z + 3| \geq 8$

   (v) $|a + 4| > 1$

   (vi) $\left|\frac{x}{2} - \frac{1}{2}\right| < 2$

3. Expand the following.

   (i) $(x-3)(x+3)$

   (ii) $(4-3x)^2$

   (iii) $(x+y)^2(x-y)$

   (iv) $(3+x)(3x+2)(x-3)$

   (v) $(x-4)^3$

4. Use Pascal's triangle (Binomial theorem) to find

   (i) the expansion of $(2+x)^4$

   (ii) the expansion of $(1+x)^8$

   (iii) the coefficient of x^5 in $(1+x)^7$.

5. Write the following expressions as partial fractions.

   (i) $\frac{3}{(x-2)(x-4)}$

   (ii) $\frac{4x-1}{(x-1)(x+2)}$

   (iii) $\frac{1}{x^2+5x+6}$

   (iv) $\frac{3x}{(x-2)(x+4)}$

   (v) $\frac{1}{(x+3)^2(x-2)}$

6. Simplify the following.

   (i) $\sqrt{27}\sqrt{3}$

   (ii) $\frac{\sqrt{5}}{\sqrt{45}}$

   (iii) $\frac{\sqrt{17}+5\sqrt{17}}{2\sqrt{17}}$

   (iv) $\frac{2}{3+\sqrt{3}}$

7. Factorise the following quadratic equations.

   (i) $y = x^2 + 6x + 5$

   (ii) $y = x^2 - 6x + 5$

   (iii) $y = x^2 + 4x - 5$

   (iv) $y = x^2 - 4x - 5$

   (v) $y = 2x^2 + x - 1$

8. Find the zeros of the following quadratics.

   (i) $y = x^2 + 4x + 4$

   (ii) $y = x^2 + 7x + 6$

   (iii) $y = x^2 + x - 12$

   (iv) $y = x^2 + x - 2$

   (v) $y = x^2 + 3x - 4$

   (vi) $y = x^2 + x - 3$

9. Use polynomial division to calculate the following.

   (i) $(x^2 + 3x + 4)/(x + 2)$

   (ii) $(x^2 + 3x + 2)/(x + 2)$

   (iii) $(x^3 + 5x^2 + 7x + 2)/(x + 2)$

10. Find the following.

    (i) $\frac{10!}{7!}$

    (ii) P_2^6

    (iii) C_2^6

    (iv) $\sum_{i=1}^{6}(i+1)$

CHAPTER 2

FUNCTIONS AND GRAPHS

2.1 THE BASIC FUNCTIONS AND CURVES

The standard functions and shapes are

1. Straight Lines: $y = mx + c$
2. Quadratics (parabolas): $y = ax^2 + bx + c$
3. Polynomials: $y = a_n x^n + \cdots + a_1 x + a_0$
4. Hyperbola: $y = \dfrac{1}{x}$
5. Exponential: $y = e^x \equiv \exp x$
6. Logarithm: $y = \ln x$
7. Sine: $y = \sin x$
8. Cosine: $y = \cos x$
9. Tangent: $y = \tan x$
10. Circles: $y^2 + x^2 = r^2$
11. Ellipses: $\left(\dfrac{y}{a}\right)^2 + \left(\dfrac{x}{b}\right)^2 = 1.$

2.2 FUNCTION PROPERTIES

A function is a rule for mapping one number to another. For example: $f(x) = x^2$ is a mapping from x to x^2 so that $f(3) = 3^2 = 9$.

EXAMPLES

1. If $f(x) = 3x + 1$ then $f(2) = 7$ and $f(a) = 3a + 1$.

2. If $f(z) = z^2 - 1$ then $f(1) = 0$.

The **domain** of a function is the set of all possible input values for that function.

EXAMPLES

1. $y = x^2 + 4$ has domain of all real numbers.

2. $y = 1/(x - 1)$ has domain $x \neq 1$. That is, all real numbers *except* $x = 1$ can be used in this function. If $x = 1$ then the function is undefined because of division by zero.

Sometimes the domain is defined as part of the function such as $y = x^2$ for $0 < x < 3$ so that the domain is restricted to be in the interval zero to three.

The **range** of a function is the set of all possible output values for that function.

EXAMPLES

1. $y = x^2$ has range $y \geq 0$ since any squared number is positive.

2. $y = \sin x$ has range $-1 \leq y \leq 1$ since the sine function is always between positive and negative one.

3. $y = x^2$, $0 < x < 3$ (so the domain is restricted to $x \in (0, 3)$) has range $0 < y < 9$.

The argument of a function could be the value of another function. For example if $f(x) = x^2$ and $g(x) = x + 1$ then

$$f(g(x)) = (g(x))^2 = (x+1)^2.$$

EXAMPLES

1. If $f(x) = 3x - 1$ then $f(x+1) = 3(x+1) - 1 = 3x + 2$.
2. If $f(x) = 2x + 1$ and $g(x) = \cos(x)$ then $f(g(x)) = 2\cos(x) + 1$ and $g(f(x)) = \cos(2x+1)$.

The **inverse** of a function is denoted $f^{-1}(x)$ and has the property that

$$f^{-1}(f(x)) = f(f^{-1}(x)) = x.$$

EXAMPLES

1. $f(x) = x^2$ and $g(x) = \sqrt{x}$ are inverses since $\sqrt{x^2} = (\sqrt{x})^2 = x$.
2. If $f(x) = 3x^2 + 1$ then the inverse is found by rearrangement:

$$\begin{aligned} f(x) &= 3x^2 + 1 \\ \Longrightarrow \quad \pm\sqrt{\frac{f(x)-1}{3}} &= x \\ \Longrightarrow \quad f^{-1}(x) &= \pm\sqrt{\frac{x-1}{3}}. \end{aligned}$$

The **zeros** of a function, $f(x)$, are the values of x when $f(x) = 0$.

EXAMPLES

1. $f(x) = 2x + 3$ has zero $x = -\frac{3}{2}$.
2. $f(x) = x^2 + 3x + 2$ has zeros $x = -1, -2$.

A graph $y = f(x)$ shifted from being centred on $(0, 0)$ to being centred on (a, b) is written in the form

$$y - b = f(x - a).$$

EXAMPLES

1. A circle with centre (1,2) has form $(x-1)^2 + (y-2)^2 = r^2$.
2. A parabola $y = x^2$ with turning point (0,0) if shifted to having turning point (3,4) has equation $(y-4) = (x-3)^2$.

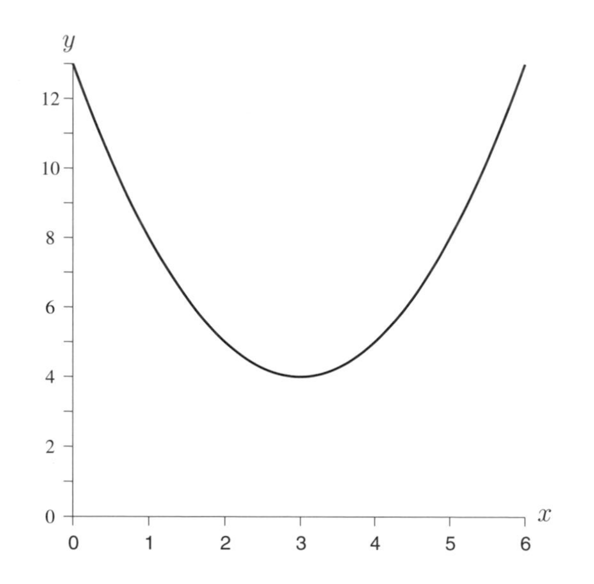

A function is **even** if $f(-x) = f(x)$ and **odd** if $f(-x) = -f(x)$.

EXAMPLES

1. $y = f(x) = x^3$ is odd since $f(-x) = (-x)^3 = -x^3 = -f(x)$.
2. $y = f(x) = x^4$ is even since $f(-x) = (-x)^4 = x^4 = f(x)$.

2.3 STRAIGHT LINES

A line has the general form

$$y = mx + a$$

where a and m are real numbers and m is the slope of the line.

EXAMPLES

1. Part of the straight line $y = 0.6 - x$ is drawn in the following diagram:

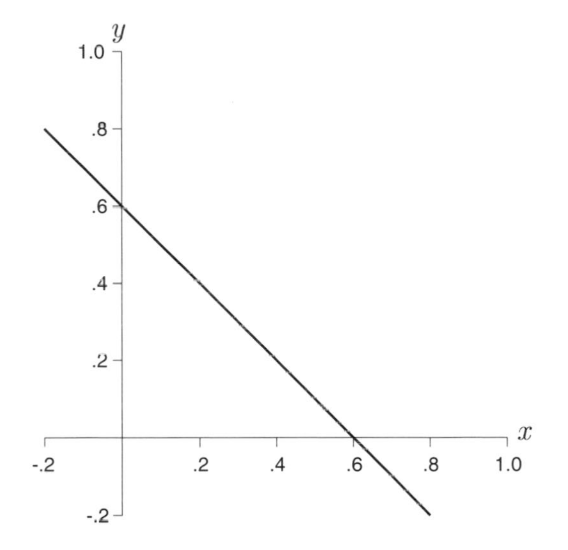

2. The line $y = 2x + 1$ cuts the x axis when $y = 0$ giving $x = -\frac{1}{2}$ as the zero.

3. The line $5y = x - 1$ has slope $m = \frac{1}{5}$ since it can be rewritten as $y = \frac{x}{5} - \frac{1}{5}$.

4. The equation of a line that passes through the points $(0, -1)$ and $(3, 0)$ is $y = \frac{x}{3} - 1$. The gradient is found from

$$m = \frac{y_2 - y_1}{x_2 - x_1} = \frac{0 + 1}{3 - 0} = \frac{1}{3}.$$

2.4 QUADRATICS

A quadratic (parabola) has the general form

$$y = ax^2 + bx + c$$

and can have either no real zeros, one real zero or two real zeros.

If the quadratic has two real zeros, c_1, c_2 then it can also be written as

$$y = a(x - c_1)(x - c_2).$$

EXAMPLE

Sections of the three quadratic functions

$$y = (x - 1)^2 + 1, \quad y = (x - 3)^2, \quad y = (x - 5)(x - 6)$$

are drawn in the following diagram:

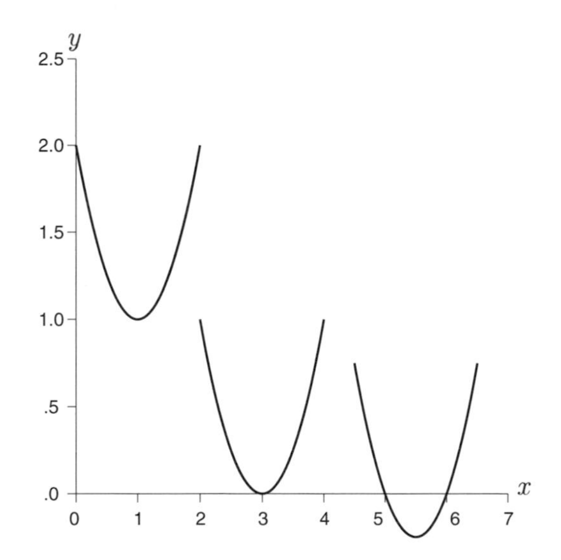

2.5 POLYNOMIALS

A polynomial has the general form

$$y = a_n x^n + a_{n-1} x^{n-1} + \cdots + a_1 x + a_0$$

where $a_i, i = 0 \ldots n$, are real numbers, and has the following properties.

1. The polynomial has **degree** n if its highest power is x^n.
2. A polynomial of degree n has n zeros (some of which may be complex).
3. The **constant** term in the above polynomial is a_0.
4. The **leading order** term in the above polynomial is $a_n x^n$ since this is the term that dominates as $x \to \infty$.

EXAMPLES

1. $y = 2x^3 + 4x^2 + 1$ has degree 3, constant term 1 and leading order term $2x^3$.
2. $y = x^2 + 5x + 6$ has two zeros $x = -3$ and $x = -2$.
3. The third degree polynomial $y = (x-1)(x-2)(x-3) = x^3 - 6x^2 + 11x - 6$ is plotted below for $x \in [0, 4]$:

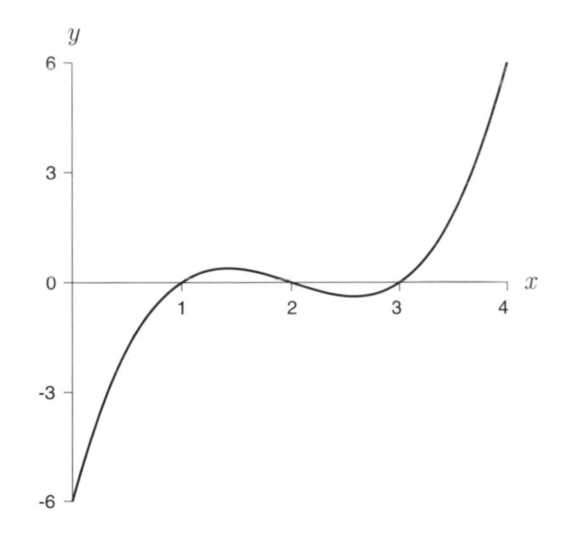

2.6 HYPERBOLA

A hyperbola centred on the origin is usually written in the form

$$y = \frac{k}{x}$$

although other orientations of hyperbolas can be written as

$$\frac{x^2}{a^2} - \frac{y^2}{b^2} = 1$$

or

$$\frac{y^2}{a^2} - \frac{x^2}{b^2} = 1.$$

EXAMPLE

The hyperbola $y = \dfrac{0.15}{x}$ is drawn in the following diagram:

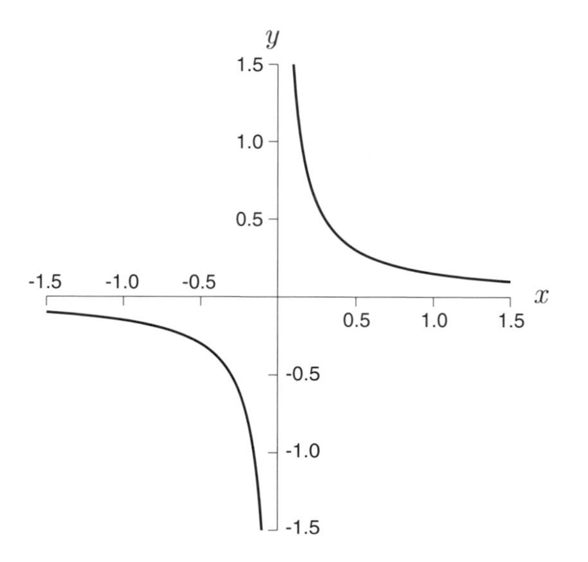

The hyperbola above is not defined for $x = 0$.

2.7 EXPONENTIAL AND LOGARITHM FUNCTIONS

The exponential function is

$$y = e^x \equiv \exp x$$

with its inverse the logarithm function

$$y = \ln x.$$

The general properties of the exponential are listed in the next chapter on transcendental functions.

EXAMPLE

The exponential function $y = e^x$ (upper curve) and logarithm function $y = \ln x$ (lower curve) are drawn in the following diagram:

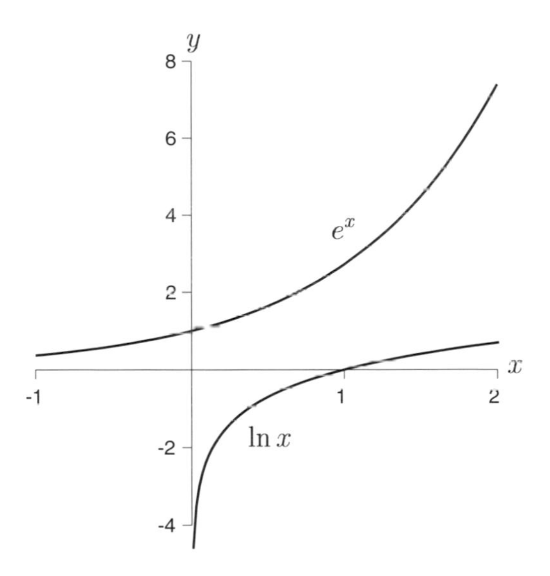

The logarithm function is not defined for $x \leq 0$.

2.8 TRIGONOMETRIC FUNCTIONS

The main trigonometric functions are $\sin x$ and $\cos x$, which are cyclic with period 2π thus $\sin(x + 2\pi) = \sin x$. Sine and Cosine can be defined in terms of angles as discussed in sections 1.14 and 3.4.

EXAMPLES

1. The function $y = \sin 2x$ will have a period of π.
2. The functions $\sin x$ and $\cos x$ are plotted below for the first period $x \in [0, 2\pi]$, while $\tan x = \sin x / \cos x$ is plotted for $x \in [-\pi/2, \pi/2]$.

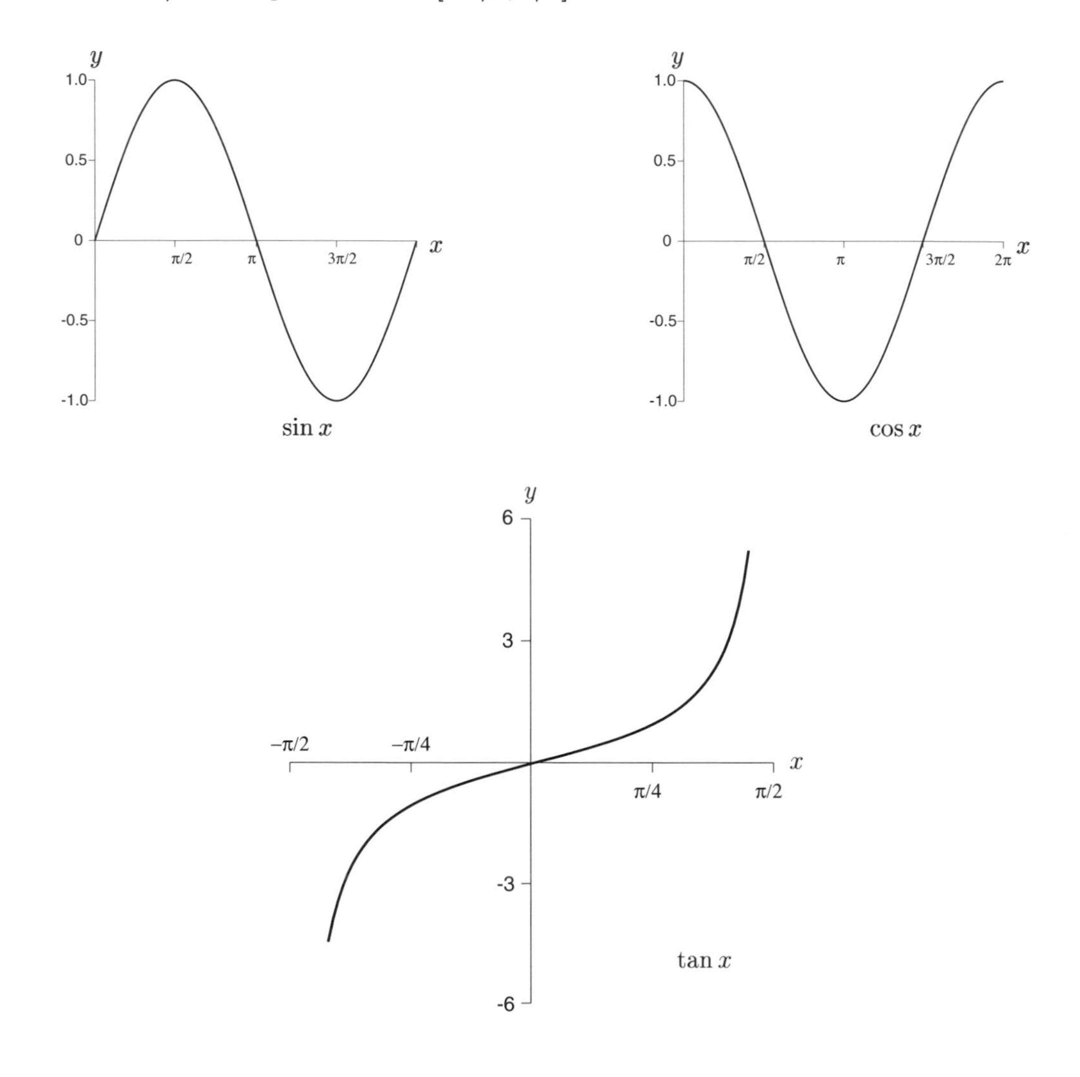

2.9 CIRCLES

A **circle** centred on the origin has the general equation

$$x^2 + y^2 = r^2$$

where r is the radius. This is often written in parametric form

$$x(t) = r\cos t, \quad y(t) = r\sin t, \quad t \in [0, 2\pi].$$

EXAMPLES

1. The circles $x^2 + y^2 = 1$ and $(x-2)^2 + (y-1.5)^2 = 1$ are drawn in the following diagram:

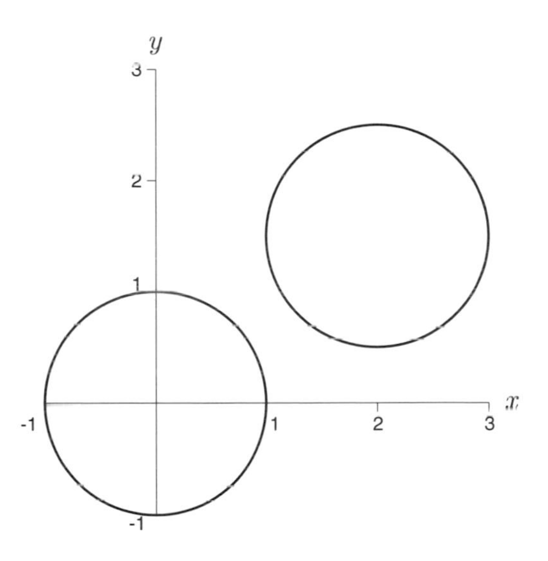

2. The curve $x^2 + 2x + y^2 + 4y = -4$ can be written as $(x+1)^2 + (y+2)^2 = 1$, which is a circle centred on $(-1, -2)$ with radius 1.

3. The curve represented by $x(t) = 2\cos t + 1$, $y(t) = 2\sin t - 3$, $t \in [0, 2\pi)$ is the circle radius 2 centred on $(1, -3)$.

2.10 ELLIPSES

An **ellipse** centred on the origin has the general equation

$$c_1x^2 + c_2xy + c_3y^2 = 1.$$

If the x and y axes are the axes of the ellipse then it is usually written in the form

$$\frac{x^2}{a^2} + \frac{y^2}{b^2} = 1$$

where $2a$ is the length of the ellipse in the x direction and $2b$ the length of the ellipse in the y direction. An ellipse is often written in parametric form

$$x(t) = a\sin t,\quad y(t) = b\cos t,\quad t \in [0, 2\pi].$$

EXAMPLES

1. The ellipse $\left(\frac{y}{2}\right)^2 + x^2 = 1$ is drawn in the following diagram:

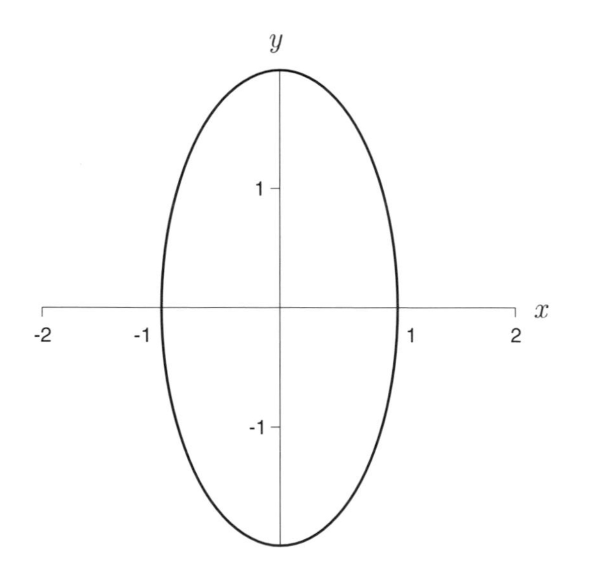

2. The curve $(x-2)^2 + 16y^2 = 1$ is an ellipse centred on $(2, 0)$ with major axis of length 2 in the x direction and minor axis of length $\frac{1}{2}$.

2.11 EXAMPLE QUESTIONS

(Answers are given in Chapter 14)

1. If $f(x) = x^3 + 1$ what is $f(2)$?
2. If $f(x) = x^3 + 1$ what is $f(g)$?
3. If $f(x) = x^3 + 1$ and $g(x) = (x - 1)$ what is $f(g(x))$?
4. If $f(x) = x^3 + 1$ and $g(x) = (x - 1)$ what is $f(g(b))$?
5. If $f(x) = x^2$ and $g(z) = \sin z$ find $f(g(a))$ and $g(f(x))$.
6. If $f(x) = x^2 + 1$ find $f(f(x))$.
7. If $f(x) = (x - 1)^2$ and $g(x) = x^2 - 1$ find $f(g(x))$ and $g(f(x))$.
8. If $f(x) = \frac{1}{x} + 1$ find the inverse $f^{-1}(x)$.
9. If $f(x) = \frac{1}{x+1}$ find the inverse $f^{-1}(x)$.
10. If $f(x) = \frac{1}{x^2} + 1$ find the inverse $f^{-1}(x)$.

Lines

11. Draw the line $y = -2x + 1$ for $x \in [0, 1]$.
12. Where is the zero of the line $y = x - 1$?
13. Where does the line $2y + x - 1 = 0$ cross the y axis? What is the slope of the line?
14. Draw $3y - x + 3 = 0$ for $x \in [0, 4]$.

Quadratics

15. Draw the quadratic $y = x^2 - 2x + 1$ for $x \in [0, 2]$.
16. Where are the zeros of the curve

$$y = (x - 3)(x - 4)?$$

(For more questions on manipulation of quadratics see Chapter 1.)

Sines and cosines

17. Draw the curve $y = 2 \sin 3x$ from $x = 0$ to $x = \pi$.
18. Draw the curve $y = \cos \frac{x}{2}$ from $x = 0$ to $x = 4\pi$.
19. Draw the curve $y = \cos 2x + 1$ from $x = 0$ to $x = 2\pi$.
20. What is the period of $y = \sin(x + 1)$?
21. What is the period of $y = \cos 3x$?
22. What is the period of $y = \sin(3x + 1)$?

Circles and ellipses

23. Draw the circle $y^2 + (x - 2)^2 = 4$.
24. Draw the ellipse $y^2 + 2x^2 = 1$.
25. Draw the ellipse $4y^2 + (x - 1)^2 = 1$.
26. Where does the ellipse $(x - 1)^2 + 2y^2 = 1$ cut the x axis?
27. What is the equation for an ellipse centred on (0,0) with x axis twice as long as the y axis?
28. What is the equation for a circle centred on $(1, 2)$ with radius 2?
29. What is the equation for a circle centred on $(a, 2)$ with radius 3?

General

30. What type of curve has equation $y^2 + (x - 1)^2 - 2 = 0$?
31. What type of curve has equation $2y^2 + (x - 1)^2 - 2 = 0$?
32. What type of curve has equation $2y + (x - 1)^2 - 2 = 0$?
33. What type of curve has equation $2y + (x - 1) = 0$?
34. What type of curve has equation $\frac{2}{y-1} + (x - 1) = 0$?
35. What is the equation of the quadratic below:

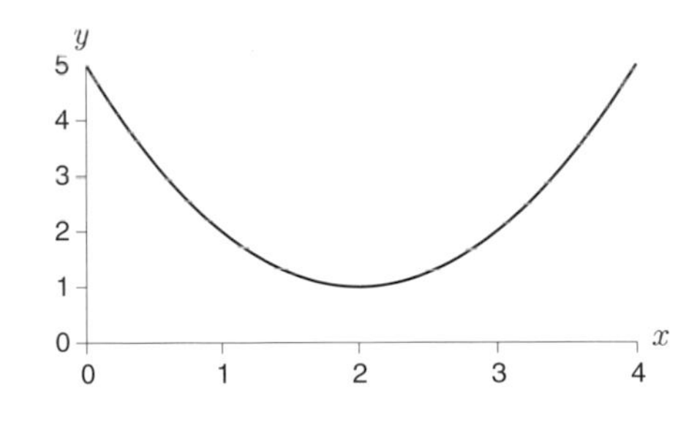

36. What is the equation of the shape below:

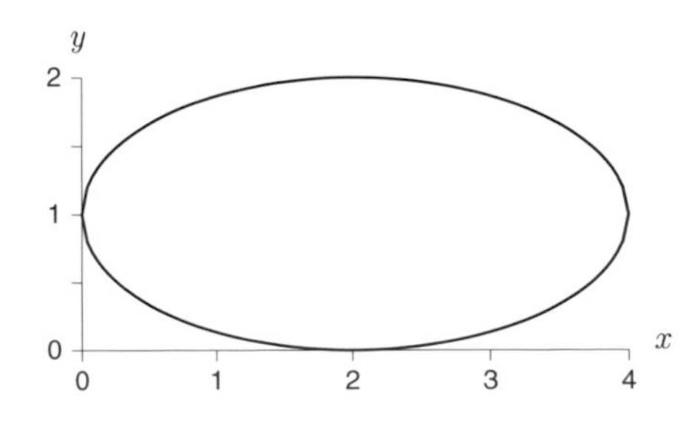

CHAPTER 3

TRANSCENDENTAL FUNCTIONS

3.1 EXPONENTIAL FUNCTION

An exponential function is defined by

$$f(x) = a^x, \qquad \text{so that} \quad x = \log_a f, \qquad a > 0,$$

where a is the **base** and x is the **index**.

EXAMPLES

1. If $8 = x^3$ then $x = 8^{1/3} = 2$.
2. If $3 = \log_2 y$ then $y = 2^3 = 8$.
3. If $2 = \log_{10} y$ then $y = 10^2 = 100$.
4. If $y = \log_2 16$ then since $16 = 2^4$, $y = 4$.

The most useful exponential function is $f(x) = e^x \equiv \exp x$ where $e = 2.71828\ldots$

3.2 INDEX LAWS

1. $a^i = \underbrace{a.a\ldots a}_{i \text{ times}}$, for i an integer.
2. $a^m a^n = a^{m+n}$ } Equal Bases Rule
3. $\dfrac{a^m}{a^n} = a^{m-n}$
4. $a^m b^m = (ab)^m$ } Equal Indices Rule
5. $\dfrac{a^m}{b^m} = \left(\dfrac{a}{b}\right)^m$
6. $a^{-n} = \dfrac{1}{a^n}$
7. $(a^m)^n = a^{mn}$ Power of a Power Rule
8. $a^0 = 1$

EXAMPLES

1. $(243)^{2/5} = ((243)^{1/5})^2 = 3^2 = 9$

2. $2^x 8^y = 2^x(2^3)^y = 2^{x+3y}$

3. To simplify $y = 3^2 9^3$ write $9 = 3^2$ so that

$$y = 3^2(3^2)^3 = 3^2 3^6 = 3^8.$$

4. If $\left(\dfrac{4}{y}\right)^3 = 64$ then

$$\frac{4}{y} = 64^{1/3} = 4$$

so $y = 1$.

3.3 LOGARITHM RULES

1. $\log_a(xy) = \log_a x + \log_a y$ — Log of a Product
2. $\log_a\left(\frac{x}{y}\right) = \log_a x - \log_a y$ — Log of a Quotient
3. $\log_a x^p = p\log_a x$ — Log of a Power
4. $\log_a(a^x) = x$
5. $a^{\log_a x} = x$
6. $\log_a 1 = 0$ and $\log_a a = 1$

EXAMPLES

1. $\log_4 16 = \log_4 4^2 = 2$
2. $9^{\log_3 x} = (3^2)^{\log_3 x} = (3^{\log_3 x})^2 = x^2$
3. If $\log a = 4, \log b = 9$ then $\log(a^2b^3) = 2\log a + 3\log b = 8 + 27 = 35$.
4. $\log_2\left(\frac{2^x}{2^y}\right) = x - y$
5. $a^{\log_a x + 2\log_a y} = a^{\log_a x}a^{\log_a y^2} = xy^2$

The natural logarithm of x, the inverse of the exponential function e^x, is $\log_e x \equiv \ln x$ (also denoted $\log x$):

$$\ln x = c \quad \text{means} \quad e^c = x.$$

Note that:

1. $\ln e^x = x$
2. $e^{\ln x} = x$
3. $\ln e = 1$
4. $\ln 1 = 0$

EXAMPLES

1. $\exp(3\ln 2) = \exp(\ln 2^3) = \exp(\ln 8) = 8$

2. $e^{x+\ln 2} = e^x e^{\ln 2} = 2e^x$

3. $e^{\ln x - 2\ln y} = \dfrac{e^{\ln x}}{e^{2\ln y}} = \dfrac{x}{y^2}$

4. If $\ln x = 2$ and $\ln y = 5$ then to find $\ln(x^3y^2)$ we write

$$\begin{aligned} \ln(x^3y^2) &= \ln x^3 + \ln y^2 \\ &= 3\ln x + 2\ln y \\ &= 3(2) + 2(5) \\ &= 16. \end{aligned}$$

5. If $\ln y = 3\ln 2x + c$ then to find y write

$$\begin{aligned} \ln y &= \ln\,(2x)^3 + c \\ \Longrightarrow \quad y &= \exp(\ln\,(2x)^3 + c) \\ &= k\exp(\ln\,(2x)^3), \quad \text{where } e^c = k \\ &= k(2x)^3. \end{aligned}$$

6. If $x = \ln 3$ and $y = \ln 4$ then to find $\exp(x + 2y)$ write

$$\begin{aligned} e^{x+2y} &= e^x(e^y)^2 \\ &= 3 \times 4^2 \\ &= 48. \end{aligned}$$

7. If $T = T_0 + T_1e^{-kt}$ then

$$k = -\frac{1}{t}\ln\left(\frac{T - T_0}{T_1}\right).$$

8. If $y = a^x$ then

$$\ln y = x\ln a \quad \Longrightarrow \quad y = e^{x\ln a}.$$

3.4 TRIGONOMETRIC FUNCTIONS

The unit circle can be used as an aid for finding the sin and cos of common angles. For example, $\cos \pi/6 = \sqrt{3}/2$. By symmetry all the other major angles can be found.

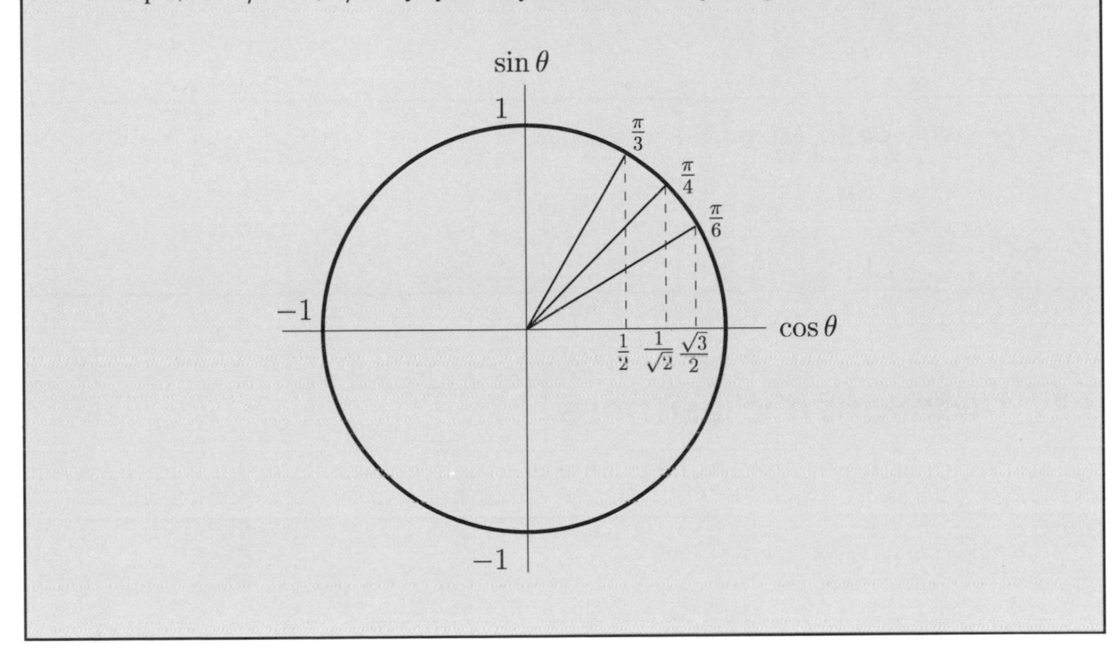

EXAMPLES

1. From the diagram we see that

$$\cos\frac{\pi}{6} = \frac{\sqrt{3}}{2}, \qquad \cos\frac{\pi}{4} = \frac{1}{\sqrt{2}}, \qquad \cos\frac{\pi}{3} = \frac{1}{2}, \qquad \cos\pi = -1.$$

2. $\sin\dfrac{5\pi}{6} \equiv \sin\dfrac{\pi}{6} = \dfrac{1}{2}$

3. $\tan\dfrac{3\pi}{4} = -\tan\dfrac{\pi}{4} = -1$

4. $\cos(n\pi) = (-1)^n, n = 0, \pm 1, \pm 2, \ldots$

5. $\sin(n\pi) = 0, n = 0, \pm 1, \pm 2, \ldots$

6. $\sin\dfrac{(2n+1)\pi}{2} = -(-1)^n, n = 0, \pm 1, \pm 2, \ldots$

$$\sin(-x) = -\sin(x), \quad \cos(-x) = \cos x$$

Sine is an odd function while cosine is an even function.

The **Reciprocal Trigonometric Functions** are

$$\operatorname{cosec} x = \frac{1}{\sin x}, \qquad \sec x = \frac{1}{\cos x}, \qquad \cot x = \frac{1}{\tan x}.$$

3.5 TRIGONOMETRIC IDENTITIES

A fundamental trigonometric identity is

$$\sin^2 x + \cos^2 x = 1.$$

EXAMPLES

1. To prove the identity $\tan x + \cot x = \sec x \operatorname{cosec} x$ consider the left hand side:

$$\begin{aligned}\tan x + \cot x &= \frac{\sin x}{\cos x} + \frac{\cos x}{\sin x} \\ &= \frac{\sin^2 x + \cos^2 x}{\cos x \sin x} \\ &= \frac{1}{\cos x \sin x} = \sec x \operatorname{cosec} x.\end{aligned}$$

2. It is easy to prove

$$\begin{aligned}1 + \tan^2 x &= \sec^2 x \\ \cot^2 x + 1 &= \operatorname{cosec}^2 x\end{aligned}$$

by simply dividing $\sin^2 x + \cos^2 x = 1$ by either $\sin^2 x$ or $\cos^2 x$.

$$\begin{aligned}
\sin(x+y) &= \sin x \cos y + \cos x \sin y \\
\cos(x+y) &= \cos x \cos y - \sin x \sin y \\
\sin 2x &= 2 \sin x \cos x \\
\cos 2x &= \cos^2 x - \sin^2 x \\
\sin^2 x &= \frac{1 - \cos 2x}{2} \\
\cos^2 x &= \frac{1 + \cos 2x}{2}
\end{aligned}$$

EXAMPLES

1. $\sin(x-y) = \sin x \cos y - \cos x \sin y$

2. $\cos(x-y) = \cos x \cos y + \sin x \sin y$

3. $\sin\left(x + \frac{\pi}{2}\right) = \sin x \, \cos \frac{\pi}{2} + \cos x \, \sin \frac{\pi}{2} = \cos x$

4. $\cos(x+\pi) = \cos x \, \cos \pi - \sin x \, \sin \pi = -\cos x$

5. To find $\sin \frac{\pi}{12}$ consider

$$\begin{aligned}
\sin \frac{\pi}{12} &= \sin\left(\frac{\pi}{3} - \frac{\pi}{4}\right) \\
&= \sin \frac{\pi}{3} \cos \frac{\pi}{4} - \cos \frac{\pi}{3} \sin \frac{\pi}{4} \\
&= \frac{\sqrt{3}}{2} \frac{1}{\sqrt{2}} - \frac{1}{2} \frac{1}{\sqrt{2}} = \frac{1}{2\sqrt{2}} \left(\sqrt{3} - 1\right).
\end{aligned}$$

6. Alternatively the following method can be used:

$$\begin{aligned}
\sin \frac{\pi}{12} &= \sqrt{\frac{1 - \cos \pi/6}{2}} \\
&= \sqrt{\frac{1 - \sqrt{3}/2}{2}}.
\end{aligned}$$

3.6 HYPERBOLIC FUNCTIONS

$$\sinh x = \frac{e^x - e^{-x}}{2}, \qquad \cosh x = \frac{e^x + e^{-x}}{2}.$$

EXAMPLES

1. It is easy to show that

$$\sinh 0 = 0$$
$$\cosh 0 = 1$$

and that

$$\cosh^2 x - \sinh^2 x = 1$$

since

$$\begin{aligned}\cosh^2 x - \sinh^2 x &= \left(\frac{e^x + e^{-x}}{2}\right)^2 - \left(\frac{e^x - e^{-x}}{2}\right)^2 \\ &= \frac{1}{4}\left[(e^{2x} + 2 + e^{-2x}) - (e^{2x} - 2 + e^{-2x})\right] = 1.\end{aligned}$$

2. The plots of $\sinh x$ and $\cosh x$ are illustrated below on the interval $x \in [-2, 2]$.

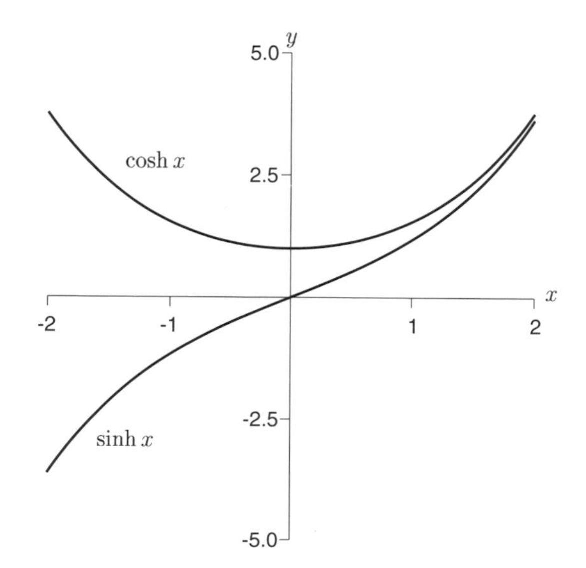

3.7 EXAMPLE QUESTIONS

(Answers are given in Chapter 14)

1. Simplify as much as possible

   (i) $6x^3y^{-2} \times \dfrac{1}{24}x^{-5}y^4$

   (ii) $8^{-\frac{2}{3}}$

   (iii) $2\log_{10} 5 + \log_{10} 8 - \log_{10} 2$

   (iv) $3^{-\log_3 P}$

   (v) $\ln x^2 + \ln y - \ln x - \ln y^2$

   (vi) $e^{2\ln x}$

2. Solve for t using natural logarithms:

   (i) $5^t = 7$

   (ii) $2 = (1.02)^t$

   (iii) $3^t 7 = 2^t 5$

   (iv) $Q = Q_0 a^{nt}$

   (v) $y = 3 - 2\ln t$

   (vi) $3y = 1 + 2e^{4t}$

3. If $\ln s = 2$ and $\ln t = 3$ calculate

   (i) $\ln(st)$

   (ii) $\ln(st^2)$

   (iii) $\ln(\sqrt{st})$

   (iv) $\ln \dfrac{s}{t}$

   (v) $\ln \dfrac{s}{t^3}$

4. If $x = \ln 3$ and $y = \ln 5$ then find

   (i) $e^x e^y$

   (ii) e^{x+y}

   (iii) e^{2x}

   (iv) $e^x + e^y$

5. Evaluate

   (i) $\tan(\pi)$

   (ii) $\sin\left(\dfrac{6\pi}{8}\right)$

   (iii) $\cos\left(\dfrac{11\pi}{6}\right)$

   (iv) $\sec\left(\dfrac{4\pi}{3}\right)$

6. Simplify

   (i) $\dfrac{1}{\cos^2\theta} - \tan^2\theta$

   (ii) $(\sin x + \cos x)^2 + (\sin x - \cos x)^2$

   (iii) $\dfrac{\tan\theta}{\sqrt{1 + \tan^2\theta}}$

7. Solve the following for values of θ between 0 and 2π

   (i) $\cos^2\theta + 3\sin^2\theta = 2$

   (ii) $2\cos^2\theta = 3\sin\theta$

8. Prove the following identities:

   (i) $\dfrac{1 + \sin\theta}{1 - \sin\theta} = (\sec\theta + \tan\theta)^2$

   (ii) $3\sin^2\theta - 2 = 1 - 3\cos^2\theta$

   (iii) $\sinh x - \cosh x = -e^{-x}$

   (iv) $\sinh x + \cosh x = e^x$

9. Use the trigonometric addition of angle formulae to show

$$\cos\frac{\pi}{12} = \frac{1}{4}(\sqrt{6} + \sqrt{2}).$$

10. For the following angles find $\cos\theta$, $\sin\theta$, $\tan\theta$, and $\sec\theta$:

    (i) $\theta = \dfrac{\pi}{4}$

    (ii) $\theta = 13\dfrac{\pi}{6}$

    (iii) $\theta = \dfrac{2\pi}{3}$

    (iv) $\theta = -\dfrac{5\pi}{3}$

    (v) $\theta = \dfrac{5\pi}{4}$

11. Use the multiple angle formulae to find $\cos\dfrac{\pi}{12}$.

12. In an experiment you have to calculate the time to melt a block of ice using the formula

$$t = \frac{l(\lambda\rho - cT_0\rho)}{hT_a}$$

    where

$$l = 0.1, \quad \lambda = 3 \times 10^5,$$
$$c = 2 \times 10^3, \quad T_0 = -20,$$
$$T_a = 20, \quad h = 10,$$
$$\rho = 1 \times 10^3.$$

    Find t.

13. Is $f(x) = x\cos x$ an odd or even function?

CHAPTER 4
DIFFERENTIATION

4.1 FIRST PRINCIPLES

The **definition of a derivative** of a function $f(x)$ is:

$$f'(x) = \frac{df}{dx} = \lim_{h \to 0} \frac{f(x+h) - f(x)}{h}.$$

This is the slope of the tangent to the function $f(x)$ at the point x. The following diagram illustrates:

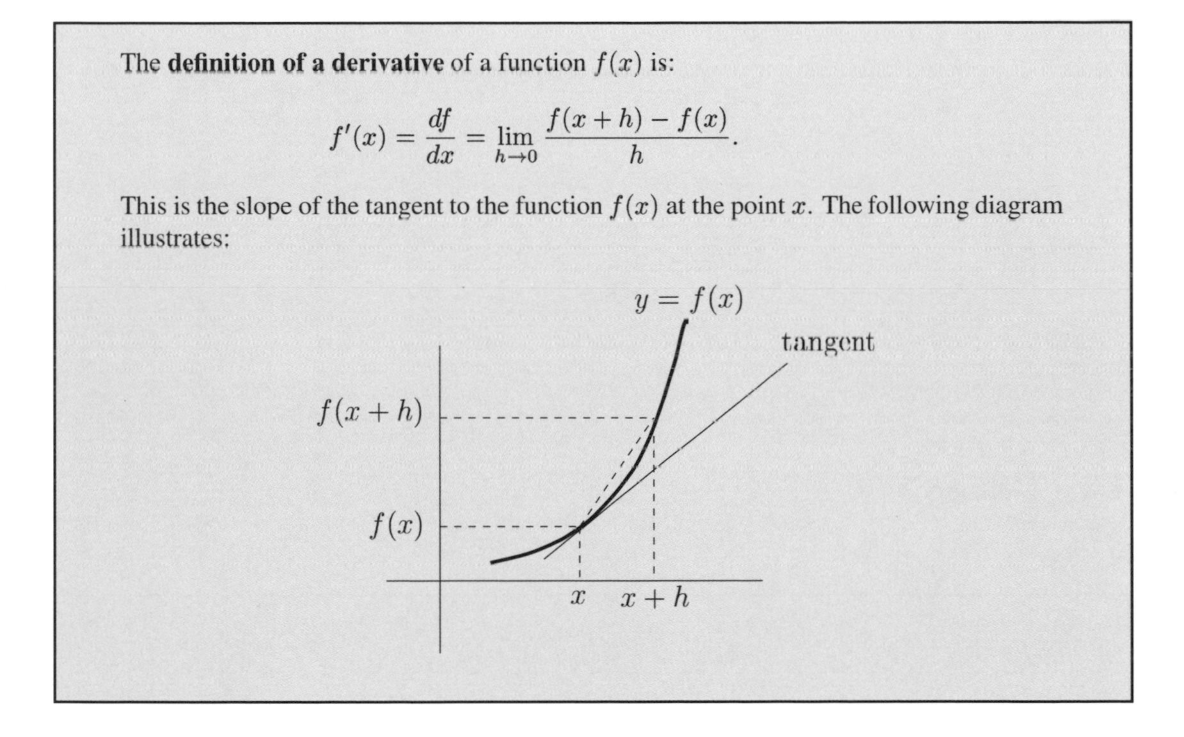

EXAMPLES

1. If $f(x) = x^2$ then

$$\begin{aligned} f'(x) &= \lim_{h\to 0} \frac{(x+h)^2 - x^2}{h} \\ &= \lim_{h\to 0} \frac{x^2 + 2hx + h^2 - x^2}{h} \\ &= \lim_{h\to 0} \frac{2hx + h^2}{h} \\ &= \lim_{h\to 0} 2x + h \\ &= 2x. \end{aligned}$$

2. If $f(x) = \sin x$ then

$$\begin{aligned} f'(x) &= \lim_{h\to 0} \frac{\sin(x+h) - \sin x}{h} \\ &= \lim_{h\to 0} \frac{\sin x \cos h + \cos x \sin h - \sin x}{h} \\ &= \lim_{h\to 0} \frac{\sin x(\cos h - 1) + \cos x \sin h}{h} \\ &= \cos x \end{aligned}$$

since

$$\lim_{h\to 0} \frac{\cos h - 1}{h} = 0, \quad \lim_{h\to 0} \frac{\sin h}{h} = 1$$

(see the Asymptotics chapter for how to evaluate these limits).

4.2 LINEARITY

$$\begin{aligned} \frac{d}{dx}(f(x) + g(x)) &= \frac{df(x)}{dx} + \frac{dg(x)}{dx} \\ \frac{d}{dx}(cf(x)) &= c\frac{df(x)}{dx} \end{aligned}$$

where c is a constant.

EXAMPLES

1. $\dfrac{d}{dx}(3\sin x) = 3\dfrac{d}{dx}\sin x = 3\cos x.$

2. If $f(x) = \sin x + e^x$ then $f'(x) = \dfrac{d}{dx}\sin x + \dfrac{d}{dx}e^x = \cos x + e^x.$

4.3 SIMPLE DERIVATIVES

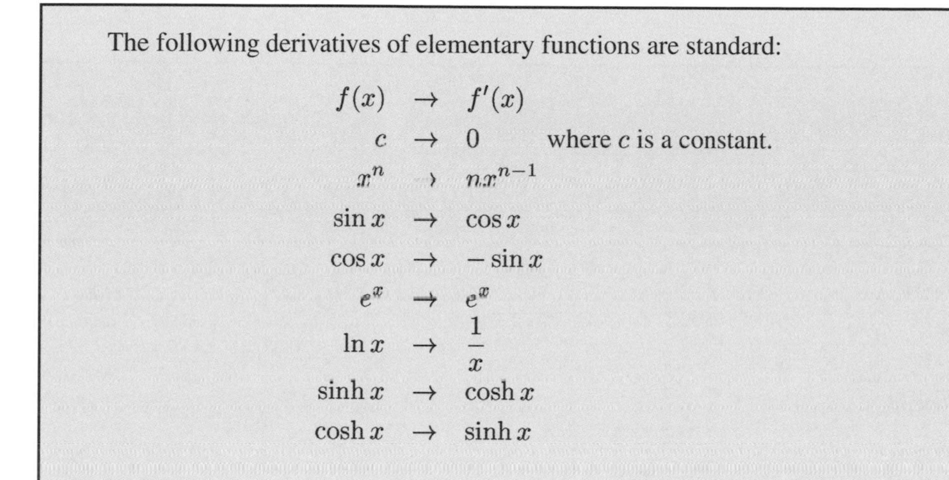

The following derivatives of elementary functions are standard:

$$\begin{array}{rcll}
f(x) & \to & f'(x) & \\
c & \to & 0 & \text{where } c \text{ is a constant.} \\
x^n & \to & nx^{n-1} & \\
\sin x & \to & \cos x & \\
\cos x & \to & -\sin x & \\
e^x & \to & e^x & \\
\ln x & \to & \dfrac{1}{x} & \\
\sinh x & \to & \cosh x & \\
\cosh x & \to & \sinh x &
\end{array}$$

EXAMPLES

1. $\dfrac{d}{dx}(4\ln x) = 4\dfrac{1}{x}$

2. If $f(x) = 5x^2 + \sinh x$ then $f'(x) = 10x + \cosh x.$

4.4 PRODUCT RULE

$$\frac{d}{dx}[f(x)\,g(x)] = \frac{df}{dx}\,g(x) + f(x)\,\frac{dg}{dx}$$

EXAMPLES

1. If $f(x) = x^2 \sin x$ then $f'(x) = 2x \sin x + x^2 \cos x$.

2. If $f(x) = \ln x \cos x$ then $f'(x) = \dfrac{1}{x} \cos x - \ln x \sin x$.

3. $\dfrac{d}{dx}(\sin x\, e^x) = \cos x\, e^x + \sin x\, e^x$

4.5 QUOTIENT RULE

$$\frac{d}{dx}\left[\frac{f(x)}{g(x)}\right] = \frac{f'(x)\,g(x) - f(x)\,g'(x)}{(g(x))^2}$$

EXAMPLES

1. If $f(x) = \dfrac{x^2}{\sin x}$ then

$$f'(x) = \frac{2x \sin x - x^2 \cos x}{\sin^2 x}.$$

2. If $f(x) = \dfrac{\sin x}{\cos x}$ then

$$f'(x) = \frac{\cos x \cos x - \sin x(-\sin x)}{\cos^2 x} = \frac{1}{\cos^2 x}.$$

Thus

$$\frac{d}{dx} \tan x = \sec^2 x.$$

3.

$$\frac{d}{dx}\left(\frac{\sin x}{x^2}\right) = \frac{\cos(x) x^2 - 2x \sin x}{x^4}$$

4.

$$\begin{aligned} \frac{d}{dx}\left(\frac{x^2 - x}{x^2 + 2}\right) &= \frac{\frac{d}{dx}[x^2 - x]\,(x^2 + 2) - (x^2 - x)\,\frac{d}{dx}[x^2 + 2]}{(x^2 + 2)^2} \\ &= \frac{(2x - 1)(x^2 + 2) - (x^2 - x)\,2x}{(x^2 + 2)^2} \\ &= \frac{x^2 + 4x - 2}{(x^2 + 2)^2} \end{aligned}$$

4.6 CHAIN RULE

$$\frac{d}{dx}\left[f(g(x))\right] = \frac{df}{dg}\frac{dg}{dx} = f'(g(x))\,g'(x)$$

Differentiate the outer function first then multiply by the derivative of the inner function.

EXAMPLES

1. Since $\frac{d}{dx}\sin x = \cos x$ then

$$\begin{aligned}\frac{d}{dx}[\sin(x^2)] &= \cos(x^2)\,\frac{d}{dx}[x^2]\\ &= \cos(x^2)2x.\end{aligned}$$

2. Since $\frac{d}{dx}\ln x = \frac{1}{x}$ then

$$\begin{aligned}\frac{d}{dx}[\ln(x + x^2)] &= \frac{1}{x + x^2}\,\frac{d}{dx}[x + x^2]\\ &= \frac{1 + 2x}{x + x^2}.\end{aligned}$$

3. Since $\frac{d}{dx}x^3 = 3x^2$ and $\frac{d}{dx}\sin x = \cos x$ then

$$\frac{d}{dx}[\sin^3 x] = 3\sin^2 x\,\cos x.$$

4. Since $\frac{d}{dx}\cos x = -\sin x$ and $\frac{d}{dx}x^5 = 5x^4$ then

$$\begin{aligned}\frac{d}{dx}[\cos((x^2 + 3x)^5)] &= -\sin((x^2 + 3x)^5)\,\frac{d}{dx}[(x^2 + 3x)^5]\\ &= -\sin((x^2 + 3x)^5)5(x^2 + 3x)^4\frac{d}{dx}[(x^2 + 3x)]\\ &= -\sin((x^2 + 3x)^5)5(x^2 + 3x)^4(2x + 3)\\ &= -5(2x + 3)(x^2 + 3x)^4\sin((x^2 + 3x)^5).\end{aligned}$$

4.7 IMPLICIT DIFFERENTIATION

To find $y'(x)$ where $y(x)$ is given implicitly, differentiate normally but treat each y as an unknown function of x. For example, if given

$$f(y) = g(x)$$

then differentiating gives

$$f'(y)\frac{dy}{dx} = g'(x) \quad \Longrightarrow \quad \frac{dy}{dx} = \frac{g'(x)}{f'(y)}$$

where the chain rule has been used to obtain the left hand side.

EXAMPLES

1. Differentiating

$$\sin y = x^2$$

with respect to x where $y = y(x)$ gives

$$\cos y \frac{dy}{dx} = 2x$$

or

$$\frac{dy}{dx} = \frac{2x}{\cos y} = \frac{2x}{\pm\sqrt{1-x^4}}$$

since $\cos y = \pm\sqrt{1-\sin^2 y} = \pm\sqrt{1-x^4}$.

2. Differentiating

$$x\cos y + y = x^3$$

with respect to x where $y = y(x)$ gives

$$\left(\cos y - x\sin(y)\frac{dy}{dx}\right) + \frac{dy}{dx} = 3x^2$$

which can be rearranged to give

$$\frac{dy}{dx} = \frac{3x^2 - \cos y}{1 - x\sin y}.$$

4.8 PARAMETRIC DIFFERENTIATION

Given $y = f(t)$ and $x = g(t)$, dy/dx may be calculated as

$$\frac{dy}{dx} = \frac{dy/dt}{dx/dt} = \frac{f'(t)}{g'(t)}.$$

EXAMPLES

1. If $y(t) = t^2$ and $x(t) = \sin t$ then

$$\frac{dy}{dx} = \frac{dy/dt}{dx/dt} = \frac{2t}{\cos t}.$$

2. If $y(t) = \sin t$ and $x(t) = \cos t$ then

$$\frac{dy}{dx} = \frac{dy/dt}{dx/dt} = \frac{\cos t}{-\sin t} = -\cot t.$$

4.9 SECOND DERIVATIVE

The second (or double) derivative is the derivative of the derivative:

$$f''(x) = \frac{d^2 f}{dx^2} = \frac{d}{dx}\left(\frac{df}{dx}\right).$$

Higher derivatives are found by repeated differentiation.

EXAMPLES

1. If $f(x) = x^4$ then $f'(x) = 4x^3$ and $f''(x) = 12x^2$.

2. If $s(t) = e^{2t}$ is the position of a particle with time t, then $s'(t) = 2e^{2t}$ is the velocity and $s''(t) = 4e^{2t}$ is the acceleration.

4.10 STATIONARY POINTS

A **stationary point** is a point (x, y) where $f'(x) = 0$. At this point the tangent to the graph is flat.

EXAMPLES

1. The function $y = x^2 + 2x + 2$ has a stationary point when

$$\frac{dy}{dx} = 2x + 2 = 0 \implies x = -1.$$

2. The function $y = 2x^3 - 9x^2 + 12x$ has stationary points when

$$\frac{dy}{dx} = 6x^2 - 18x + 12 = 0 \implies x = 1,\ 2.$$

3. The function $y = xe^{-x}$ has a maximum when

$$\frac{dy}{dx} = e^{-x}(x - 1) = 0 \implies x = 1.$$

A **local maximum** is when the function at the stationary point is higher than the surrounding points. A **local minimum** is lower than the surrounding points. An **inflection** point is where the graph is flat but neither a maximum nor minimum.

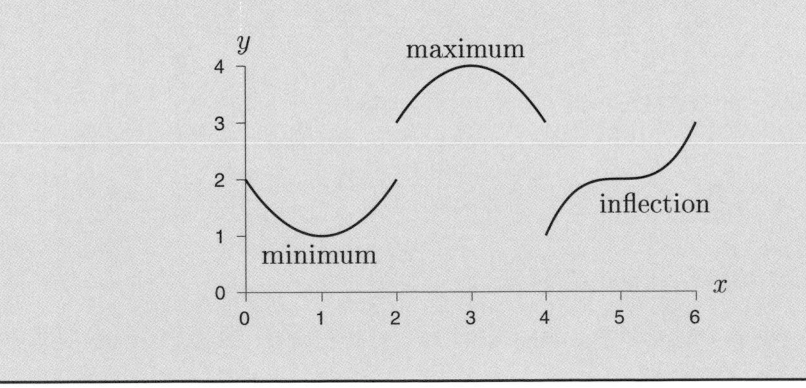

At a stationary point $x = a$ the **second derivative** indicates the type of stationary point:

1. if $f''(a) > 0$ then $x = a$ is a local minimum.
2. if $f''(a) < 0$ then $x = a$ is a local maximum.
3. if $f''(a) = 0$ then $x = a$ is an inflection point.

Note that $x = a$ is a stationary point so $f'(a) = 0$.

EXAMPLES

1. The function $y = x^2 + 2x + 2$ has a stationary point at $x = -1$. The double derivative is

$$\frac{d^2y}{dx^2} = 2$$

so $x = -1$ is a minimum.

2. The function $y = 2x^3 - 9x^2 + 12x$ has stationary points at $x = 1$ and $x = 2$. The double derivative is

$$\frac{d^2y}{dx^2} = 12x - 18$$

which is positive at $x = 2$ (a minimum) and negative at $x = 1$ (a maximum).

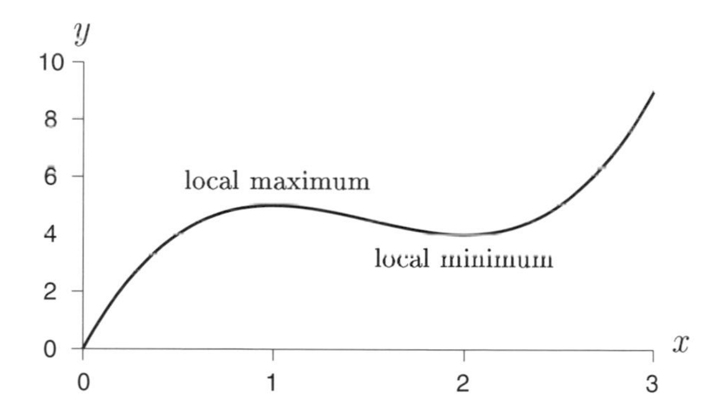

3. The function $y = (x - 1)^3 + 3$ has derivatives

$$\frac{dy}{dx} = 3(x - 1)^2, \quad \frac{d^2y}{dx^2} = 6(x - 1)$$

which are both zero at $x = 1$, which is therefore an inflection point.

4.11 EXAMPLE QUESTIONS

(Answers are given in Chapter 14)

1. Use *linearity* to find dy/dx:

   (i) $y = 3\sin x - 5\cos x$

   (ii) $y = 3e^x - x^2$

   (iii) $y = 3\ln x$

   (iv) $y = 2\sinh x - 3\cosh x$

2. Use the *chain rule* to find dy/dx:

   (i) $y = \sin(2x)$

   (ii) $y = \sin(x + x^3)$

   (iii) $y = (x+4)^3$

   (iv) $y = (x + \sin x)^5$

   (v) $y = \sin(\ln x^2)$

   (vi) $y = \exp(\cos^2 x)$

   (vii) $y = \cosh(2x^2)$

3. Use the *product* rule or the *quotient* rule to find dy/dx:

   (i) $y = xe^x$

   (ii) $y = \dfrac{\cos x}{x^2}$

   (iii) $y = e^x \sin x$

   (iv) $y = \dfrac{\ln x}{x^4}$

   (v) $y = \sin x \cos x$

   (vi) $y = \dfrac{\ln x}{e^x}$

   (vii) $y = x^2 \sin x$

   (viii) $y = \cosh x \sinh x$

   (ix) $y = \dfrac{e^{1/x}}{x}$

4. Find dy/dx for these more difficult problems:

   (i) $y = \exp(x\cos x^2)$

   (ii) $y = e^x \cos((2x+1)^2)$

   (iii) $y = \dfrac{1}{\sqrt{2+x^2}}$

   (iv) $y = \dfrac{\sin x}{(x+1)^2}$

   (v) $y = \sin(x^2 + \exp(x^3 + x))$

   (vi) $y = \dfrac{\exp x^2}{x^2}$

5. Use *implicit* differentiation to find dy/dx:

   (i) $y^2 = \sin(x-1)$

   (ii) $\cos(2y) = (1-x^2)^{1/2}$

   (iii) $\ln(y) = xe^x$

   (iv) $e^y = e^{3x} + 5$

   (v) $y + y^3 = x^2$

   (vi) $y^2 + \sin y = \sin x$

   (vii) $y(x+1) - y^2 = x$

6. Use *parametric* differentiation to find dy/dx:

   (i) $y(t) = \cos t, \quad x(t) = \sin(t^2)$

   (ii) $y(t) = e^t, \quad x(t) = t^2$

   (iii) $y(t) = t^2, \quad x(t) = \sin t$

7. Find the derivative dT/dt:

$$T = t \exp\left(\frac{a}{\sqrt{\pi t}}\right)$$

   where a is a constant.

8. For the following functions find the stationary points and classify them.

   (i) $y = (x-2)^2$

   (ii) $y = x^3 - 6x^2 + 9x + 1$

   (iii) $y = 3x^4 - 8x^3 + 6x^2$

   (iv) $y = xe^{-x}$

   (v) $y = x^2 \ln(x)$

   (vi) $y = \sin x + (1-x)\cos x$, for $x \in [-1, 2]$

   (vii) $y = (x-1)^2 e^x$

9. The function $y = f(x)$ is drawn below. Roughly sketch the function $f'(x)$.

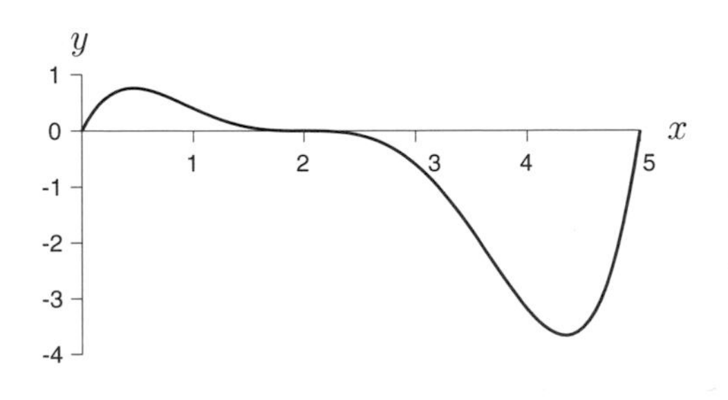

CHAPTER 5

INTEGRATION

5.1 ANTIDIFFERENTIATION

The **indefinite integral** (antiderivative) of f with respect to x is

$$\int f(x)\,dx = F(x) + c$$

where $F'(x) = f(x)$ and c is known as the **constant of integration**.

EXAMPLES

1. If $\dfrac{d}{dx}x^2 = 2x$ then $\displaystyle\int 2x\,dx = x^2 + c.$

2. If $\dfrac{d}{dx}\ln x = \dfrac{1}{x}$ then $\displaystyle\int \frac{1}{x}\,dx = \ln|x| + c.$

3. If $\dfrac{d}{dx}\sin x = \cos x$ then $\displaystyle\int \cos x\,dx = \sin x + c.$

4. If $\dfrac{d}{dx}\sinh x = \cosh x$ then $\displaystyle\int \cosh x\,dx = \sinh x + c.$

5.2 SIMPLE INTEGRALS

The following integrals of elementary functions are standard:

$$\begin{aligned}
\int x^n\,dx &= \frac{1}{n+1}x^{n+1}+c \qquad \text{where } n \neq -1\\
\int \cos x\,dx &= \sin x + c\\
\int \sin x\,dx &= -\cos x + c\\
\int e^x\,dx &= e^x + c\\
\int \frac{1}{x}\,dx &= \ln|x| + c\\
\int \sinh x\,dx &= \cosh x + c\\
\int \cosh x\,dx &= \sinh x + c
\end{aligned}$$

Integration is **linear** so that

$$\begin{aligned}
\int (f(x)+g(x))\,dx &= \int f(x)\,dx + \int g(x)\,dx,\\
\int cf(x)\,dx &= c\int f(x)\,dx, \qquad \text{where } c \text{ is a constant.}
\end{aligned}$$

EXAMPLES

1. $\int (\sin x + e^x)\,dx = -\cos x + e^x + c$

2. $\int 5\cos x\,dx = 5\sin x + c$

3. Simple application of the Chain Rule in differentiation gives

$$\int \cos kx\,dx = \frac{1}{k}\sin kx + c.$$

4. $\int 6x\,dx = 3\int 2x\,dx = 3x^2 + c$

5. $\int 4x + \frac{1}{x}\,dx = 2\int 2x\,dx + \int \frac{1}{x}\,dx = 2x^2 + \ln x + c$

6. $\int 3\sin(2x)\,dx = -\frac{3}{2}\cos(2x) + c$

7. $\int (5x^4 + \frac{1}{2}x^3 + 12x^2 + 7)\,dx = x^5 + \frac{1}{8}x^4 + 4x^3 + 7x + c$

8. $\int \left(\frac{5}{x} - e^{x/3}\right)\,dx = 5\ln|x| - 3e^{x/3} + c$

5.3 THE DEFINITE INTEGRAL

The **definite integral** with respect to x over the interval $[a, b]$ is written as:

$$\begin{aligned}\int_a^b f(x)\,dx &= [F(x)]_a^b \\ &= F(b) - F(a)\end{aligned}$$

where $F'(x) = f(x)$. This is the **Fundamental Theorem of Calculus.**

EXAMPLES

1. $\int_1^2 x^2\,dx = \left[\frac{1}{3}x^3\right]_1^2 = \frac{2^3}{3} - \frac{1}{3} = \frac{7}{3}$

2. $\int_0^1 \sin \pi x\,dx = \left[-\frac{1}{\pi}\cos \pi x\right]_0^1 = -\frac{1}{\pi}(\cos\pi - \cos 0) = \frac{2}{\pi}$

3. $\int_1^3 x^3\,dx = \left[\frac{x^4}{4}\right]_1^3 = \frac{81}{4} - \frac{1}{4} = 20$

4. $\int_{-1}^1 \sinh x\,dx = \cosh(1) - \cosh(-1) = 0$ since $\cosh(x)$ is even.

$$\int_a^a f(x)\,dx = 0$$

$$\int_a^b f(x)\,dx = -\int_b^a f(x)\,dx$$

$$\int_a^b f(x)\,dx + \int_b^c f(x)\,dx = \int_a^c f(x)\,dx$$

(assuming $f(x)$ can be integrated over the required intervals).

EXAMPLES

1.

$$\begin{aligned}\int_{-1}^{2} |x|\,dx &= \int_{-1}^{0} |x|\,dx + \int_{0}^{2} |x|\,dx \\ &= \int_{-1}^{0} -x\,dx + \int_{0}^{2} x\,dx \\ &= -\left[\frac{x^2}{2}\right]_{-1}^{0} + \left[\frac{x^2}{2}\right]_{0}^{2} \\ &= -\left[0 - \frac{1}{2}\right] + [2 - 0] = \frac{5}{2}\end{aligned}$$

2. If $f(x) = \begin{cases} 1, & x < 1, \\ x, & x \geq 1 \end{cases}$ then

$$\begin{aligned}\int_{0}^{3} f(x)\,dx &= \int_{0}^{1} f(x)\,dx + \int_{1}^{3} f(x)\,dx \\ &= \int_{0}^{1} 1\,dx + \int_{1}^{3} x\,dx \\ &= [x]_0^1 + \left[\frac{x^2}{2}\right]_1^3 \\ &= 1 + \frac{9}{2} - \frac{1}{2} = 5.\end{aligned}$$

3. If $f(x)$ is odd then

$$\int_{-a}^{a} f(x)\,dx = \int_{-a}^{0} f(x)\,dx + \int_{0}^{a} f(x)\,dx = 0.$$

5.4 AREAS

If f is an integrable function then

$$\int_a^b f(x)\,dx = \text{(area above the } x\text{-axis) - (area below the } x\text{-axis)}$$

in the region $a \leq x \leq b$.

EXAMPLES

1. Consider the curve given by $f(x) = x^3 - 9x^2 + 26x - 24 = (x-2)(x-3)(x-4)$. The area between the curve and the x-axis between $x = 2$ and $x = 4$ is given by:

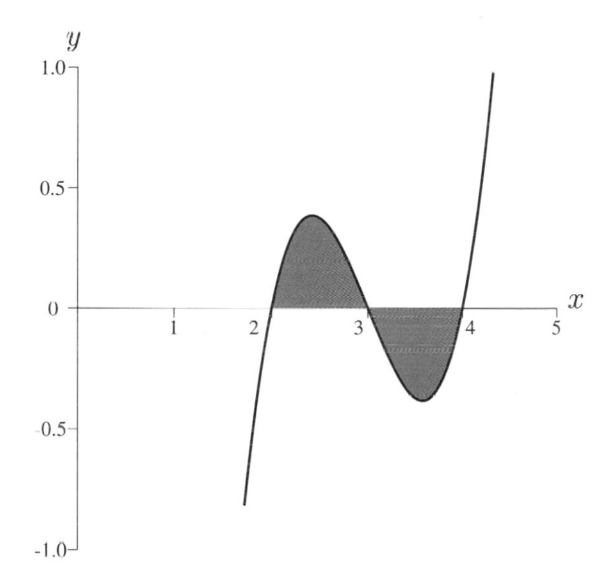

$$\begin{aligned}\int_2^4 |f(x)|\,dx &= \int_2^3 f(x)\,dx - \int_3^4 f(x)\,dx \\ &= 0.25 + 0.25 \\ &= 0.5.\end{aligned}$$

2. If $f(x)$ is an even function (so $f(-x) = f(x)$) then

$$\int_{-c}^{c} f(x)\,dx = 2\int_0^c f(x)\,dx$$

since the area for $x \in [-c, 0]$ is the same as for $x \in [0, c]$.

5.5 INTEGRATION BY SUBSTITUTION

Integrals that can be written in the form

$$\int f(g(x))g'(x)\,dx$$

are solved by the substitution $u = g(x)$, upon which the integral becomes

$$\int f(u)\,du = F(u) + c = F(g(x)) + c,$$

where $F'(x) = f(x)$. For definite integrals the limits of the integration are also transformed.

EXAMPLES

1. To evaluate $\int 2x\sin(x^2+1)\,dx$ let $u = x^2+1$ then $\frac{du}{dx} = 2x$ so that

$$\begin{aligned}\int 2x\sin(x^2+1)\,dx &= \int \sin u\,du \\ &= -\cos u + c \\ &= -\cos(x^2+1) + c.\end{aligned}$$

2. To evaluate $\int \frac{1}{\sqrt{x+1}}\,dx$ let $u = x+1$ then $\frac{du}{dx} = 1$ so that

$$\begin{aligned}\int \frac{1}{\sqrt{x+1}}\,dx &= \int u^{-\frac{1}{2}}\,du \\ &= 2u^{\frac{1}{2}} + c \\ &= 2\sqrt{x+1} + c.\end{aligned}$$

3. To find $\int_0^{\pi/2} \sin^4 x\cos x\,dx$ let $u = \sin x$ so that the integral becomes

$$\int_{\sin(0)}^{\sin(\pi/2)} u^4\,du = \left[\frac{u^5}{5}\right]_0^1 = \frac{1}{5}.$$

4. To find $\int \frac{1}{\sqrt{1-x^2}}\,dx$ let $x = \sin u$ since $\sqrt{1-x^2} = \cos u$ and $dx = \cos u\,du$ so that

$$\int \frac{1}{\sqrt{1-x^2}}\,dx = \int \frac{\cos u}{\cos u}\,du = u = \arcsin x + c.$$

5.6 INTEGRATION BY PARTS

Integration of a product of two functions can sometimes be solved by integration by parts:

$$\int u\frac{dv}{dx}\,dx = uv - \int v\frac{du}{dx}\,dx$$

or in short hand,

$$\int u\,dv = uv - \int v\,du.$$

EXAMPLES

1. To evaluate $\int x\cos x\,dx$ let $u = x$ and $\frac{dv}{dx} = \cos x$ then $\frac{du}{dx} = 1$ and $v = \sin x$, so that

$$\begin{aligned}\int x\cos x\,dx &= x\sin x - \int \sin x\,dx \\ &= x\sin x - (-\cos x + c_1) \\ &= x\sin x + \cos x + c_2.\end{aligned}$$

2. The integral $\int_0^1 xe^{2x}\,dx$ is performed by setting $u = x$, $\frac{dv}{dx} = e^{2x}$ so that $\frac{du}{dx} = 1$ and $v = \frac{1}{2}e^{2x}$. Then we have

$$\begin{aligned}\int_0^1 xe^{2x}\,dx &= \left[x\frac{1}{2}e^{2x}\right]_0^1 - \int_0^1 \frac{1}{2}e^{2x}\,dx \\ &= \frac{1}{2}e^2 - \frac{1}{4}\left[e^{2x}\right]_0^1 = \frac{1}{4}(e^2+1).\end{aligned}$$

3. Integrating by parts twice we can evaluate

$$\begin{aligned}\int e^{2x}\sin x\,dx &= -e^{2x}\cos x + 2\int e^{2x}\cos x\,dx \\ &= -e^{2x}\cos x + 2\left(e^{2x}\sin x - 2\int e^{2x}\sin x\,dx\right)\end{aligned}$$

so by rearranging $\int e^{2x}\sin x\,dx = \frac{e^{2x}}{5}(2\sin x - \cos x) + c$.

5.7 EXAMPLE QUESTIONS

(Answers are given in Chapter 14)

1. Find

(i) $\int \frac{7}{x^3}\,dx$

(ii) $\int \left(\frac{12}{x} + x^5\right)\,dx$

(iii) $\int (e^{7x} + e^{-14x})\,dx$

(iv) $\int (x^9 + x^{10} + x^{11} + x^{12})\,dx$

(v) $\int \sinh(2x)\,dx$

(vi) $\int 4\cosh(x) - e^x\,dx$

2. Evaluate

(i) $\int_0^5 e^{-z}\,dz$

(ii) $\int_{\pi/2}^{3\pi/2} \sin x\,dx$

(iii) $\int_{1/4}^{1/2} \frac{16}{x}\,dx$

(iv) $\int_0^{\pi} \sin(3x)\,dx$

(v) $\int_{\pi/4}^{5\pi/4} -\sin\left(-\frac{x}{3}\right)\,dx$

(vi) $\int_0^3 7e^{-x}\,dx$

3. Find

(i) $\int_0^2 f(x)\,dx$ where

$$f(x) = \begin{cases} 1, & x < 1, \\ x, & x \geq 1. \end{cases}$$

(ii) $\int_{-1}^1 f(x)\,dx$ where

$$f(x) = \begin{cases} -x^2, & x < 0, \\ x^3, & x \geq 0. \end{cases}$$

(iii) $\int_0^{\infty} f(x)\,dx$ where

$$f(x) = \begin{cases} x - 1, & x < 1, \\ 0, & x \geq 1. \end{cases}$$

(iv) $\int_{-1}^1 f(x)\,dx$ where $f(x) = |x^3|$.

4. Evaluate the following integrals using a substitution.

(i) $\int xe^{-x^2}\,dx$

(ii) $\int \frac{x+2}{x^2+4x+5}\,dx$

(iii) $\int \frac{1}{y\ln y}\,dy$

(iv) $\int 2x^3\sqrt{7x^4 - 1}\,dx$

(v) $\int z\cos(-z^2)\,dz$

(vi) $\int \frac{e^{\sqrt{s}}}{\sqrt{s}}\,ds$

(vii) $\int_0^{\pi} x\sin(x^2)\,dx$

(viii) $\int_9^4 \frac{e^{-\sqrt{y}}}{2\sqrt{y}}\,dy$

5. Evaluate the following integrals using integration by parts.

(i) $\int_0^{\pi} x\cos x\,dx$

(ii) $\int_0^2 \frac{1}{x^2 - 2x - 3}\,dx$

(iii) $\int e^{2x}\sin x\,dx$ (integrate twice).

(iv) $\int (x+1)\sin x\,dx$

(v) $\int x^2 e^x\,dx$

(vi) $\int \ln x\,dx$ by using $u = \ln x$ and $dv = 1$.

6. Find the following integrals using any method.

(i) $\int \frac{\sin z}{\cos z}\,dz$

(ii) $\int x^2 \sin x\,dx$

(iii) $\int 3xe^{x/3}\,dx$

(iv) $\int \sqrt{u}(u+1)\,du$

(v) $\int \frac{1}{\sqrt{1+x^2}}\,dx$ (let $x = \sinh u$).

CHAPTER 6
MATRICES

6.1 ADDITION

If $\mathbf{A}$ and $\mathbf{B}$ are $m \times n$ matrices such that

$$\mathbf{A} = \begin{bmatrix} a_{11} & a_{12} & \cdots & a_{1n} \\ a_{21} & a_{22} & \cdots & a_{2n} \\ \vdots & \vdots & \ddots & \vdots \\ a_{m1} & a_{m2} & \cdots & a_{mn} \end{bmatrix} \quad \text{and} \quad \mathbf{B} = \begin{bmatrix} b_{11} & b_{12} & \cdots & b_{1n} \\ b_{21} & b_{22} & \cdots & b_{2n} \\ \vdots & \vdots & \ddots & \vdots \\ b_{m1} & b_{m2} & \cdots & b_{mn} \end{bmatrix}$$

$$\text{then} \quad \mathbf{A} + \mathbf{B} = \begin{bmatrix} a_{11}+b_{11} & a_{12}+b_{12} & \cdots & a_{1n}+b_{1n} \\ a_{21}+b_{21} & a_{22}+b_{22} & \cdots & a_{2n}+b_{2n} \\ \vdots & \vdots & \ddots & \vdots \\ a_{m1}+b_{m1} & a_{m2}+b_{m2} & \cdots & a_{mn}+b_{mn} \end{bmatrix}.$$

Addition of matrices of different sizes is not defined.

EXAMPLES

1.
$$\begin{bmatrix} 12 & 3 & 4 \\ -2 & 1 & 0 \\ 3 & 7 & 9 \end{bmatrix} + \begin{bmatrix} -6 & 2 & -3 \\ 0 & -1 & 7 \\ 3 & -14 & 1 \end{bmatrix} = \begin{bmatrix} 6 & 5 & 1 \\ -2 & 0 & 7 \\ 6 & -7 & 10 \end{bmatrix}$$

2.
$$\begin{bmatrix} 1 & 2 \\ 2 & 1 \end{bmatrix} + \begin{bmatrix} 1 \\ 2 \end{bmatrix} \quad \text{cannot be done.}$$

6.2 MULTIPLICATION

AB is **defined** if **A** is size $m \times r$ and **B** size $r \times n$. If

$$\mathbf{A} = \begin{bmatrix} a_{11} & a_{12} & \cdots & a_{1m} \\ a_{21} & a_{22} & \cdots & a_{2m} \\ \vdots & \vdots & \ddots & \vdots \\ a_{r1} & a_{r2} & \cdots & a_{rm} \end{bmatrix} \quad \text{and} \quad \mathbf{B} = \begin{bmatrix} b_{11} & b_{12} & \cdots & b_{1r} \\ b_{21} & b_{22} & \cdots & b_{2r} \\ \vdots & \vdots & \ddots & \vdots \\ b_{n1} & b_{n2} & \cdots & b_{nr} \end{bmatrix}$$

then $\mathbf{AB} = \mathbf{C}$ is an $m \times n$ matrix, where $C_{ij} = a_{i1}b_{1j} + a_{i2}b_{2j} + \cdots + a_{ir}b_{rj}$. That is, C_{ij} is the dot product of row i of **A** and column j of **B**.

In general $\mathbf{AB} \neq \mathbf{BA}$, that is, matrices are **noncommutative**.

EXAMPLES

1. $\begin{bmatrix} 1 & 2 \\ 3 & 4 \end{bmatrix} \begin{bmatrix} 1 & -1 \\ 1 & 2 \end{bmatrix} = \begin{bmatrix} 3 & 3 \\ 7 & 5 \end{bmatrix}$

2. $\begin{bmatrix} 1 & -1 \\ 1 & 2 \end{bmatrix} \begin{bmatrix} 1 & 2 \\ 3 & 4 \end{bmatrix} = \begin{bmatrix} -2 & -2 \\ 7 & 10 \end{bmatrix}$

3. If $\mathbf{A} = \begin{bmatrix} 4 & 2 & 9 \\ -3 & -1 & 1 \\ 2 & 1 & 2 \end{bmatrix}$ and $\mathbf{B} = \begin{bmatrix} -2 & 2 & -4 \\ 0 & 0 & 1 \\ 3 & -4 & -1 \end{bmatrix}$ then

$$\begin{aligned} \mathbf{AB} &= \begin{bmatrix} 4 & 2 & 9 \\ -3 & -1 & 1 \\ 2 & 1 & 2 \end{bmatrix} \begin{bmatrix} -2 & 2 & -4 \\ 0 & 0 & 1 \\ 3 & -4 & -1 \end{bmatrix} \\ &= \begin{bmatrix} -8+0+27 & 8+0-36 & -16+2-9 \\ 6+0+3 & -6+0-4 & 12-1-1 \\ -4+0+6 & 4+0-8 & -8+1-2 \end{bmatrix} \\ &= \begin{bmatrix} 19 & -28 & -23 \\ 9 & -10 & 10 \\ 2 & -4 & -9 \end{bmatrix}, \end{aligned}$$

while

$$
\begin{aligned}
\mathbf{BA} &= \begin{bmatrix} -2 & 2 & -4 \\ 0 & 0 & 1 \\ 3 & -4 & -1 \end{bmatrix} \begin{bmatrix} 4 & 2 & 9 \\ -3 & -1 & 1 \\ 2 & 1 & 2 \end{bmatrix} \\
&= \begin{bmatrix} -8-6-8 & -4-2-4 & -18+2-8 \\ 0+0+2 & 0+0+1 & 0+0+2 \\ 12+12-2 & 6+4-1 & 27-4-2 \end{bmatrix} \\
&= \begin{bmatrix} -22 & -10 & -24 \\ 2 & 1 & 2 \\ 22 & 9 & 21 \end{bmatrix} \\
&\neq \mathbf{AB}.
\end{aligned}
$$

4.

$$
\begin{bmatrix} 1 & 2 \\ 1 & 1 \\ 1 & 0 \end{bmatrix} \begin{bmatrix} 1 \\ 3 \end{bmatrix} = \begin{bmatrix} 7 \\ 4 \\ 1 \end{bmatrix}
$$

6.3 IDENTITY

The **identity matrix**, defined only for square matrices ($n \times n$), is

$$
\mathbf{I} = \begin{bmatrix} 1 & 0 & \cdots & 0 \\ 0 & 1 & \cdots & 0 \\ \vdots & \vdots & \ddots & \vdots \\ 0 & 0 & \cdots & 1 \end{bmatrix}
$$

and is defined such that, for all $n \times n$ matrices $\mathbf{A}$,

$$
\mathbf{IA} = \mathbf{AI} = \mathbf{A}.
$$

EXAMPLE

The 2×2 and 3×3 identity matrices are

$$
\begin{bmatrix} 1 & 0 \\ 0 & 1 \end{bmatrix}, \quad \begin{bmatrix} 1 & 0 & 0 \\ 0 & 1 & 0 \\ 0 & 0 & 1 \end{bmatrix}.
$$

6.4 TRANSPOSE

The transpose of a matrix is formed by writing its columns as rows. The transpose of an $m \times n$ matrix $\mathbf{A}$ is an $n \times m$ matrix denoted by $\mathbf{A}^t$, that is, if

$$\mathbf{A} = \begin{bmatrix} a_{11} & a_{12} & \cdots & a_{1n} \\ a_{21} & a_{22} & \cdots & a_{2n} \\ \vdots & \vdots & \ddots & \vdots \\ a_{m1} & a_{m2} & \cdots & a_{mn} \end{bmatrix} \quad \text{then} \quad \mathbf{A}^t = \begin{bmatrix} a_{11} & a_{21} & \cdots & a_{m1} \\ a_{12} & a_{22} & \cdots & a_{m2} \\ \vdots & \vdots & \ddots & \vdots \\ a_{1n} & a_{2n} & \cdots & a_{mn} \end{bmatrix}.$$

EXAMPLES

1. If $\mathbf{A} = \begin{bmatrix} 0 & 1 \\ 2 & 4 \\ 1 & -1 \end{bmatrix}$ then $\mathbf{A}^t = \begin{bmatrix} 0 & 2 & 1 \\ 1 & 4 & -1 \end{bmatrix}$.

2. If $\mathbf{A} = \begin{bmatrix} 4 & 2 & 9 \\ -3 & -1 & 1 \\ 2 & 1 & 2 \end{bmatrix}$ then $\mathbf{A}^t = \begin{bmatrix} 4 & -3 & 2 \\ 2 & -1 & 1 \\ 9 & 1 & 2 \end{bmatrix}$.

If $\mathbf{A}$ and $\mathbf{B}$ are matrices and c is a scalar, then

1. $(\mathbf{A}^t)^t = \mathbf{A}$
2. $(\mathbf{A}+\mathbf{B})^t = \mathbf{A}^t + \mathbf{B}^t$
3. $(c\mathbf{A})^t = c\mathbf{A}^t$
4. $(\mathbf{AB})^t = \mathbf{B}^t\mathbf{A}^t$

EXAMPLE

$(A^tA)^t = (A^t(A^t)^t) = A^tA$

6.5 DETERMINANTS

The determinant of a 2×2 matrix $\mathbf{A} = \begin{bmatrix} a & b \\ c & d \end{bmatrix}$ is

$$\det(\mathbf{A}) = |\mathbf{A}| = ad - bc.$$

The determinant of a 3×3 matrix $\mathbf{A} = \begin{bmatrix} a_{11} & a_{12} & a_{13} \\ a_{21} & a_{22} & a_{23} \\ a_{31} & a_{32} & a_{33} \end{bmatrix}$ is

$$|\mathbf{A}| = a_{11}\begin{vmatrix} a_{22} & a_{23} \\ a_{32} & a_{33} \end{vmatrix} - a_{12}\begin{vmatrix} a_{21} & a_{23} \\ a_{31} & a_{33} \end{vmatrix} + a_{13}\begin{vmatrix} a_{21} & a_{22} \\ a_{31} & a_{32} \end{vmatrix}$$

(expanding by the first row).

EXAMPLES

1.

$$\begin{vmatrix} 1 & 2 \\ 3 & 4 \end{vmatrix} = 4 - 6 = -2, \qquad \begin{vmatrix} 1 & 2 \\ 2 & -1 \end{vmatrix} = -1 - 4 = -5.$$

2.

$$\begin{vmatrix} -3 & 2 & 1 \\ 4 & 5 & 6 \\ 2 & -3 & 1 \end{vmatrix} = -3\begin{vmatrix} 5 & 6 \\ -3 & 1 \end{vmatrix} - 2\begin{vmatrix} 4 & 6 \\ 2 & 1 \end{vmatrix} + 1\begin{vmatrix} 4 & 5 \\ 2 & -3 \end{vmatrix}$$
$$= -3(5 + 18) - 2(4 - 12) + (-12 - 10)$$
$$= -75$$

3.

$$\begin{vmatrix} 1 & 7 & 5 \\ 0 & 2 & 6 \\ 0 & 0 & 3 \end{vmatrix} = 1\begin{vmatrix} 2 & 6 \\ 0 & 3 \end{vmatrix} + 0 + 0$$
$$= 1 \times 2 \times 3 = 6$$

4.

$$\begin{vmatrix} 1 & 7 & 5 \\ 0 & 2 & 6 \end{vmatrix}$$ is not possible.

6.5.1 COFACTOR EXPANSION

The determinant of an $n \times n$ matrix may be found by choosing a row (or column) and summing the products of the entries of the chosen row (or column) and their cofactors:

$$\det(\mathbf{A}) = a_{1j}C_{1j} + a_{2j}C_{2j} + \cdots + a_{nj}C_{nj},$$

(cofactor expansion along the jth column)

$$\det(\mathbf{A}) = a_{i1}C_{i1} + a_{i2}C_{i2} + \cdots + a_{in}C_{in},$$

(cofactor expansion along the ith row)

where C_{ij} is the determinant of $\mathbf{A}$ with row i and column j deleted, multiplied by $(-1)^{i+j}$. The matrix of elements C_{ij} is called the **cofactors matrix**.

EXAMPLES

1.
$$\begin{vmatrix} 1 & 3 & 0 & 2 \\ 4 & -3 & 1 & 9 \\ -4 & 4 & 0 & 3 \\ 5 & -5 & -2 & -7 \end{vmatrix}$$

(Expansion is along the 3rd column since it has two zeros.)

$$\begin{aligned} &= (0)C_{13} + (1)C_{23} + (0)C_{33} + (-2)C_{43} \\ &= (1)(-1)^5 \begin{vmatrix} 1 & 3 & 2 \\ -4 & 4 & 3 \\ 5 & -5 & -7 \end{vmatrix} + (-2)(-1)^7 \begin{vmatrix} 1 & 3 & 2 \\ 4 & -3 & 9 \\ -4 & 4 & 3 \end{vmatrix} \\ &= -[1(-28+15) - 3(28-15) + 2(20-20)] \\ &\quad + 2[1(-9-36) - 3(12+36) + 2(16-12)] \\ &= -[13-39] + 2[-45-144+8] \\ &= -310. \end{aligned}$$

2.

$$\begin{vmatrix} 1 & 2 & 3 \\ 0 & 0 & 2 \\ 1 & 1 & 0 \end{vmatrix} = -2 \begin{vmatrix} 1 & 2 \\ 1 & 1 \end{vmatrix} = -2(1-2) = 2$$

by expanding along the second row.

3. The full cofactors matrix for the previous question is found by crossing out each row and column

in turn rcmcmbcring to multiply by $(-1)^{i+j}$:

$$C_{11} = +1\begin{vmatrix} 0 & 2 \\ 1 & 0 \end{vmatrix} = -2$$

$$C_{12} = (-1)\begin{vmatrix} 0 & 2 \\ 1 & 0 \end{vmatrix} = 2$$

$$C_{13} = +1\begin{vmatrix} 0 & 0 \\ 1 & 1 \end{vmatrix} = 0$$

and so one, giving

$$\mathbf{C} = \begin{bmatrix} -2 & 2 & 0 \\ 3 & -3 & 1 \\ 4 & -2 & 0 \end{bmatrix}.$$

6.6 INVERSE

A square matrix $\mathbf{A}$ is said to be **invertible** if there exists $\mathbf{B}$ such that

$$\mathbf{AB} = \mathbf{BA} = \mathbf{I}.$$

$\mathbf{B}$ is denoted $\mathbf{A}^{-1}$ and is unique.

If $\det(\mathbf{A}) = 0$ then a matrix is not invertible.

EXAMPLE

The matrix $\mathbf{B} = \begin{bmatrix} 3 & 5 \\ 1 & 2 \end{bmatrix}$ is the inverse of $\mathbf{A} = \begin{bmatrix} 2 & -5 \\ -1 & 3 \end{bmatrix}$ since

$$\mathbf{AB} = \begin{bmatrix} 3 & 5 \\ 1 & 2 \end{bmatrix}\begin{bmatrix} 2 & -5 \\ -1 & 3 \end{bmatrix} = \begin{bmatrix} 1 & 0 \\ 0 & 1 \end{bmatrix} = \mathbf{I}$$

and

$$\mathbf{BA} = \begin{bmatrix} 2 & -5 \\ -1 & 3 \end{bmatrix}\begin{bmatrix} 3 & 5 \\ 1 & 2 \end{bmatrix} = \begin{bmatrix} 1 & 0 \\ 0 & 1 \end{bmatrix} = \mathbf{I}.$$

6.6.1 TWO BY TWO MATRICES

For 2×2 matrices, if $\mathbf{A} = \begin{bmatrix} a & b \\ c & d \end{bmatrix}$ then

$$\mathbf{A}^{-1} = \frac{1}{ad - bc} \begin{bmatrix} d & -b \\ -c & a \end{bmatrix} \text{ providing } ad - bc \neq 0.$$

If $\det(\mathbf{A}) = ad - bc = 0$ then $\mathbf{A}^{-1}$ does not exist.

EXAMPLES

1. If $\mathbf{A} = \begin{bmatrix} 1 & 2 \\ 3 & 4 \end{bmatrix}$ then $\mathbf{A}^{-1} = \frac{1}{-2} \begin{bmatrix} 4 & -2 \\ -3 & 1 \end{bmatrix}$.

2. If $\mathbf{A} = \begin{bmatrix} 1 & 2 \\ 0 & 3 \end{bmatrix}$ then $\mathbf{A}^{-1} = \frac{1}{3} \begin{bmatrix} 3 & -2 \\ 0 & 1 \end{bmatrix}$.

6.6.2 PARTITIONED MATRIX

Inverses can also be found by considering the partitioned matrix

$$\left[\mathbf{A} \;\vdots\; \mathbf{I} \right]$$

then performing row operations until the final partitioned matrix is of the form

$$\left[\mathbf{I} \;\vdots\; \mathbf{A}^{-1} \right].$$

EXAMPLE

The inverse of

$$\begin{bmatrix} 1 & 2 & 1 \\ 0 & 1 & 0 \\ 1 & 1 & 0 \end{bmatrix}$$

can be calculated using row reductions where $R3 \to R3 - R1$ means that Row 3 becomes the old

Row 3 minus Row 1.

$$\left[\begin{array}{ccc|ccc} 1 & 2 & 1 & 1 & 0 & 0 \\ 0 & 1 & 0 & 0 & 1 & 0 \\ 1 & 1 & 0 & 0 & 0 & 1 \end{array}\right]$$

$$\downarrow \quad R3 \to R3 - R1$$

$$\left[\begin{array}{ccc|ccc} 1 & 2 & 1 & 1 & 0 & 0 \\ 0 & 1 & 0 & 0 & 1 & 0 \\ 0 & -1 & -1 & -1 & 0 & 1 \end{array}\right]$$

$$\downarrow \quad R3 \to R3 + R2$$

$$\left[\begin{array}{ccc|ccc} 1 & 2 & 1 & 1 & 0 & 0 \\ 0 & 1 & 0 & 0 & 1 & 0 \\ 0 & 0 & -1 & -1 & 1 & 1 \end{array}\right]$$

$$\downarrow \quad \begin{array}{l} R1 \to R1 - 2R2 \\ R3 \to -R3 \end{array}$$

$$\left[\begin{array}{ccc|ccc} 1 & 0 & 1 & 1 & -2 & 0 \\ 0 & 1 & 0 & 0 & 1 & 0 \\ 0 & 0 & 1 & 1 & -1 & -1 \end{array}\right]$$

$$\downarrow \quad R1 \to R1 - R3$$

$$\left[\begin{array}{ccc|ccc} 1 & 0 & 0 & 0 & -1 & 1 \\ 0 & 1 & 0 & 0 & 1 & 0 \\ 0 & 0 & 1 & 1 & -1 & -1 \end{array}\right]$$

hence

$$\begin{bmatrix} 1 & 2 & 1 \\ 0 & 1 & 0 \\ 1 & 1 & 0 \end{bmatrix}^{-1} = \begin{bmatrix} 0 & -1 & 1 \\ 0 & 1 & 0 \\ 1 & -1 & -1 \end{bmatrix}.$$

6.6.3 COFACTORS MATRIX

The inverse of a $n \times n$ matrix $\mathbf{A}$ can be found by considering the transpose of the cofactors matrix divided by the determinant:

$$\mathbf{A}^{-1} = \frac{1}{|\mathbf{A}|}\mathbf{C}^t$$

where C_{ij} is the determinant of $\mathbf{A}$ with row i and column j deleted, multiplied by $(-1)^{i+j}$. The matrix $\mathbf{C}$ is called the cofactors matrix.

EXAMPLES

1. If

$$\mathbf{A} = \begin{bmatrix} 1 & 2 & 1 \\ 0 & 1 & 0 \\ 1 & 1 & 0 \end{bmatrix}$$

then

$$C_{11} = \begin{vmatrix} 1 & 0 \\ 1 & 0 \end{vmatrix} = 0$$

$$C_{21} = (-1)\begin{vmatrix} 2 & 1 \\ 1 & 0 \end{vmatrix} = 1$$

and so on. Since $|\mathbf{A}| = -1$ we get

$$\mathbf{A}^{-1} = \frac{1}{-1}\begin{bmatrix} 0 & 0 & -1 \\ 1 & -1 & 1 \\ -1 & 0 & -1 \end{bmatrix}^T = \begin{bmatrix} 0 & -1 & 1 \\ 0 & 1 & 0 \\ 1 & -1 & -1 \end{bmatrix}.$$

2. The matrix

$$\mathbf{A} = \begin{bmatrix} 1 & 2 & 3 \\ 0 & 0 & 2 \\ 1 & 1 & 0 \end{bmatrix}$$

has cofactors matrix

$$\mathbf{C} = \begin{bmatrix} -2 & 2 & 0 \\ 3 & -3 & 1 \\ 4 & -2 & 0 \end{bmatrix}$$

hence the inverse

$$\mathbf{A}^{-1} = \frac{1}{2}\begin{bmatrix} -2 & 3 & 4 \\ 2 & -3 & -2 \\ 0 & 1 & 0 \end{bmatrix}.$$

6.7 MATRIX MANIPULATION

Matrices do not behave as real numbers. When manipulating matrix expressions a distinction is made between multiplying from the left (pre-multiplication) and multiplying from the right (post-multiplication).

EXAMPLES

1. Given that $\mathbf{ABC} = \mathbf{I}$ find $\mathbf{B}$?

$$
\begin{aligned}
\mathbf{ABC} &= \mathbf{I} \\
(\mathbf{A}^{-1}\mathbf{A})\,\mathbf{BC} &= \mathbf{A}^{-1}\mathbf{I} && \text{pre-multiply both sides by } \mathbf{A}^{-1} \\
I\,\mathbf{B}\,(\mathbf{CC}^{-1}) &= \mathbf{A}^{-1}\mathbf{IC}^{-1} && \text{post-multiply both sides by } \mathbf{C}^{-1} \\
\mathbf{B} &= \mathbf{A}^{-1}\mathbf{C}^{-1} && \text{simplifying} \\
&= (\mathbf{CA})^{-1}.
\end{aligned}
$$

2. If $\mathbf{A} = \mathbf{PDP}^{-1}$, then $\mathbf{A}^3$ is

$$
\begin{aligned}
\mathbf{A}^3 &= (\mathbf{PDP}^{-1})\,(\mathbf{PDP}^{-1})\,(\mathbf{PDP}^{-1}) \\
&= \mathbf{PD}\,(\mathbf{P}^{-1}\mathbf{P})\,\mathbf{DP}^{-1}\mathbf{PDP}^{-1} \\
&= \mathbf{PD}^2\mathbf{P}^{-1}\mathbf{PDP}^{-1} && \text{since } \mathbf{PP}^{-1} = \mathbf{I} \\
&= \mathbf{PD}^3\mathbf{P}^{-1} && \text{again since } \mathbf{PP}^{-1} = \mathbf{I}.
\end{aligned}
$$

3. If $\mathbf{A}\underset{\sim}{\boldsymbol{v}} = \lambda\underset{\sim}{\boldsymbol{v}}$ then

$$
\begin{aligned}
\mathbf{A}^3\underset{\sim}{\boldsymbol{v}} &= \mathbf{AA}\,\mathbf{A}\underset{\sim}{\boldsymbol{v}} = \mathbf{AA}(\lambda\underset{\sim}{\boldsymbol{v}}) = \lambda\mathbf{AA}\underset{\sim}{\boldsymbol{v}} \\
&= \lambda^2\mathbf{A}\underset{\sim}{\boldsymbol{v}} \\
&= \lambda^3\underset{\sim}{\boldsymbol{v}}.
\end{aligned}
$$

4. If $\mathbf{A}\underset{\sim}{\boldsymbol{v}} = \lambda\underset{\sim}{\boldsymbol{v}}$ then if $\mathbf{A}^{-1}$ exists then

$$
\begin{aligned}
&\mathbf{A}^{-1}\,\mathbf{A}\underset{\sim}{\boldsymbol{v}} = \mathbf{A}^{-1}\,\lambda\underset{\sim}{\boldsymbol{v}} = \lambda\mathbf{A}^{-1}\underset{\sim}{\boldsymbol{v}} \\
\Longrightarrow\quad &\underset{\sim}{\boldsymbol{v}} = \lambda\mathbf{A}^{-1}\underset{\sim}{\boldsymbol{v}} \\
\Longrightarrow\quad &\frac{1}{\lambda}\underset{\sim}{\boldsymbol{v}} = \mathbf{A}^{-1}\underset{\sim}{\boldsymbol{v}}.
\end{aligned}
$$

See Section 6.9 on eigenvalues since this example shows that if $\mathbf{A}$ has eigenvalue λ, with eigenvector $\underset{\sim}{\boldsymbol{v}}$, then $\mathbf{A}^{-1}$ has eigenvalue $1/\lambda$ for the same eigenvector.

6.8 SYSTEMS OF EQUATIONS

Systems of m linear equations involving n unknowns may be written as a matrix equation. For example,

$$\begin{aligned} x + y + 2z &= 1 \\ 2x + 4y - 3z &= 5 \\ 3x + 6y - 5z &= 2 \end{aligned}$$

is written as

$$\begin{bmatrix} 1 & 1 & 2 \\ 2 & 4 & -3 \\ 3 & 6 & -5 \end{bmatrix} \begin{bmatrix} x \\ y \\ z \end{bmatrix} = \begin{bmatrix} 1 \\ 5 \\ 2 \end{bmatrix}$$

or

$$\mathbf{Ax} = \mathbf{b}.$$

Systems of equations are typically solved by **Gaussian elimination**.

If $\mathbf{A}$ is invertible then $\mathbf{x} = \mathbf{A}^{-1}\mathbf{b}$.

Gaussian Elimination allows

- a multiple of one row to be added to another row.
- a row to be multiplied by a (non-zero) number.

Hence $R3 \to R3 - 2R1$ means each element in Row 3 becomes the old Row 3 element minus two times the corresponding Row 2 element.

EXAMPLES

1. The **augmented matrix** is an easy way of writing systems of equations. For the following system

$$\begin{aligned} x + y + 2z &= 1 \\ 2x + 4y - 3z &= 5 \\ 3x + 6y - 5z &= 2 \end{aligned}$$

the augmented matrix is

$$\left[\begin{array}{ccc|c} 1 & 1 & 2 & 1 \\ 2 & 4 & -3 & 5 \\ 3 & 6 & -5 & 2 \end{array}\right]$$

$$\downarrow \quad \begin{array}{l} R_2 \to R_2 - 2R_1 \\ R_3 \to R_3 - 3R_1 \end{array}$$

$$\left[\begin{array}{ccc|c} 1 & 1 & 2 & 1 \\ 0 & 2 & -7 & 3 \\ 0 & 3 & -11 & -1 \end{array}\right]$$

$$\downarrow \quad R_2 \to R_2/2$$

$$\left[\begin{array}{ccc|c} 1 & 1 & 2 & 1 \\ 0 & 1 & -7/2 & 3/2 \\ 0 & 3 & -11 & -1 \end{array}\right]$$

$$\downarrow \quad R_3 \to R_3 - 3R_2$$

$$\left[\begin{array}{ccc|c} 1 & 1 & 2 & 1 \\ 0 & 1 & -7/2 & 3/2 \\ 0 & 0 & -1/2 & -11/2 \end{array}\right]$$

$$\downarrow \quad R_3 \to -2R_3$$

$$\left[\begin{array}{ccc|c} 1 & 1 & 2 & 1 \\ 0 & 1 & -7/2 & 3/2 \\ 0 & 0 & 1 & 11 \end{array}\right]$$

This gives the straightforward solution by back substitution of $x = -61, y = 40, z = 11$.

2. Consider the system

$$\begin{aligned} 2x - 5y &= -2 \\ -x + 3y &= 4 \end{aligned}$$

written as $\mathbf{Ax} = \mathbf{b}$ such that

$$\mathbf{A} = \begin{bmatrix} 2 & -5 \\ -1 & 3 \end{bmatrix}, \quad \mathbf{x} = \begin{bmatrix} x \\ y \end{bmatrix}, \quad \mathbf{b} = \begin{bmatrix} -2 \\ 4 \end{bmatrix}.$$

The matrix $\mathbf{A}$ has inverse $\mathbf{A}^{-1} = \begin{bmatrix} 3 & 5 \\ 1 & 2 \end{bmatrix}$ so

$$\begin{aligned} \begin{bmatrix} x \\ y \end{bmatrix} &= \begin{bmatrix} 3 & 5 \\ 1 & 2 \end{bmatrix} \begin{bmatrix} -2 \\ 4 \end{bmatrix} \\ &= \begin{bmatrix} 14 \\ 6 \end{bmatrix}. \end{aligned}$$

After performing Gaussian reduction by row operations the three cases (no solution, infinite solutions, one solution) are typically represented by the following:

1. If you perform row operations to obtain

$$\left[\begin{array}{ccc|c} a & b & c & k_1 \\ 0 & d & e & k_2 \\ 0 & 0 & f & k_3 \end{array}\right]$$

   (where $a, .., f$ are *non-zero* real numbers) then you get **one** unique solution.

2. If you perform row operations to obtain

$$\left[\begin{array}{ccc|c} a & b & c & k_1 \\ 0 & d & e & k_2 \\ 0 & 0 & 0 & k_3 \end{array}\right]$$

   then if $k_3 \neq 0$ you get **no** solution.

3. If you perform row operations to obtain

$$\left[\begin{array}{ccc|c} a & b & c & k_1 \\ 0 & d & e & k_2 \\ 0 & 0 & 0 & 0 \end{array}\right]$$

   then you get **an infinite** number of solutions that represent a line where you let $z = t$, t is some parameter, and then express x, y in terms of t.

EXAMPLE

To solve the system:

$$\begin{bmatrix} 1 & -2 & -1 \\ 2 & 1 & 3 \\ 1 & 8 & 9 \end{bmatrix} \begin{bmatrix} x \\ y \\ z \end{bmatrix} = \begin{bmatrix} -1 \\ 13 \\ 29 \end{bmatrix}$$

perform row reductions to obtain

$$\left[\begin{array}{ccc|c} 1 & -2 & -1 & -1 \\ 0 & 1 & 1 & 3 \\ 0 & 0 & 0 & 0 \end{array}\right]$$

and setting $z = t$ gives $y = 3 - t$ and $x - 2(3 - t) - t = -1$ so

$$(x, y, z) = (5 - t, 3 - t, t) = (5, 3, 0) + t(-1, -1, 1)$$

or a line in three dimensional space.

6.9 EIGENVALUES AND EIGENVECTORS

If $\mathbf{A}$ is an $n \times n$ matrix then a scalar λ is called an **eigenvalue** of $\mathbf{A}$, if associated with it there is a non-zero vector $\underset{\sim}{\boldsymbol{v}}$, called an **eigenvector**, such that

$$\mathbf{A}\underset{\sim}{\boldsymbol{v}} = \lambda \underset{\sim}{\boldsymbol{v}}.$$

To find the eigenvalues solve the characteristic equation

$$|\mathbf{A} - \lambda \mathbf{I}| = 0.$$

To find the eigenvectors solve

$$(\mathbf{A} - \lambda \mathbf{I})\underset{\sim}{\boldsymbol{v}} = \underset{\sim}{\mathbf{0}}.$$

EXAMPLE

To find the eigenvalues and eigenvectors for

$$\mathbf{A} = \begin{bmatrix} 0 & 1 \\ 1 & 0 \end{bmatrix}$$

set up the characteristic equation

$$\det\left(\begin{bmatrix} 0 & 1 \\ 1 & 0 \end{bmatrix} - \lambda \begin{bmatrix} 1 & 0 \\ 0 & 1 \end{bmatrix}\right) = \begin{vmatrix} -\lambda & 1 \\ 1 & -\lambda \end{vmatrix} = 0\,,$$

which gives $\lambda^2 - 1 = 0$ so $\lambda_1 = 1$ and $\lambda_2 = -1$ are the eigenvalues. To find the eigenvectors solve

$$\begin{bmatrix} -\lambda & 1 \\ 1 & -\lambda \end{bmatrix} \underset{\sim}{\boldsymbol{v}} = \mathbf{0}.$$

For $\lambda_1 = 1$ let

$$\underset{\sim}{\boldsymbol{v_1}} = \begin{bmatrix} x_1 \\ y_1 \end{bmatrix} \quad \text{then} \quad \begin{bmatrix} -1 & 1 \\ 1 & -1 \end{bmatrix} \begin{bmatrix} x_1 \\ y_1 \end{bmatrix} = \begin{bmatrix} 0 \\ 0 \end{bmatrix}.$$

Both equations give $x_1 - y_1 = 0$ so y_1 is a free variable. Hence the eigenvector corresponding to $\lambda_1 = 1$ is $t(1, 1)$, where t is any number, $t \in \text{Re}$.
For $\lambda_2 = -1$ let

$$\underset{\sim}{\boldsymbol{v_2}} = \begin{bmatrix} x_2 \\ y_2 \end{bmatrix} \quad \text{then} \begin{bmatrix} 1 & 1 \\ 1 & 1 \end{bmatrix} \begin{bmatrix} x_2 \\ y_2 \end{bmatrix} = \begin{bmatrix} 0 \\ 0 \end{bmatrix}.$$

Both equations give $x_2 + y_2 = 0$ so y_2 is a free variable. Hence the eigenvector corresponding to $\lambda_2 = -1$ is $p(1, -1)$. The length of the eigenvector is unimportant hence it is convenient to write

$$\underset{\sim}{\boldsymbol{v_1}} = (1, 1), \qquad \underset{\sim}{\boldsymbol{v_2}} = (1, -1).$$

6.10 TRACE

> The **trace** of a matrix is the sum of its diagonal elements.
> (Note that the trace is also equal to the sum of the eigenvalues.)

EXAMPLE The trace of $\begin{bmatrix} 1 & 2 & 3 \\ 4 & 5 & 6 \\ 7 & 8 & 9 \end{bmatrix} = 1 + 5 + 9 = 15.$

6.11 SYMMETRIC MATRICES

> The matrix **A** is **symmetric** if $\mathbf{A} = \mathbf{A}^t$.

EXAMPLE The matrix $\begin{bmatrix} 1 & 2 & 3 \\ 2 & 5 & 6 \\ 3 & 6 & 9 \end{bmatrix}$ is symmetric.

6.12 DIAGONAL MATRICES

> A **diagonal** matrix is one with only terms along the main diagonal.

EXAMPLE A 3×3 diagonal matrix has the form $\begin{bmatrix} a & 0 & 0 \\ 0 & b & 0 \\ 0 & 0 & c \end{bmatrix}$.

6.13 EXAMPLE QUESTIONS

(Answers are given in Chapter 14)

1. Find $\mathbf{A}+\mathbf{B}$, $\mathbf{AB}$, $\mathbf{BA}$ and the trace($\mathbf{A}$):

(i)

$$\mathbf{A} = \begin{bmatrix} 4 & 0 & 1 \\ 2 & 3 & -7 \\ 1 & 0 & 0 \end{bmatrix}$$

$$\mathbf{B} = \begin{bmatrix} 1 & -1 & -1 \\ -2 & -2 & -2 \\ 3 & 1 & 7 \end{bmatrix}$$

(ii)

$$\mathbf{A} = \begin{bmatrix} 1 & 9 \\ 6 & 4 \end{bmatrix}$$

$$\mathbf{B} = \begin{bmatrix} 6 & -8 \\ -7 & 1 \end{bmatrix}$$

(iii)

$$\mathbf{A} = \begin{bmatrix} 2 & 0 & 0 \\ 0 & 3 & 0 \\ 0 & 0 & 4 \end{bmatrix}$$

$$\mathbf{B} = \begin{bmatrix} 9 & 0 & 0 \\ 0 & 8 & 0 \\ 0 & 0 & -3 \end{bmatrix}$$

2. Find $\mathbf{AB}$:

(i) $\mathbf{A} = \begin{bmatrix} 2 & 9 \\ 1 & 3 \\ 4 & 2 \end{bmatrix}$ $\mathbf{B} = \begin{bmatrix} 6 & 1 \\ 3 & 1 \end{bmatrix}$

(ii) $\mathbf{A} = \begin{bmatrix} 1 & -1 & 1 & -1 \end{bmatrix}$

$\mathbf{B} - \begin{bmatrix} 2 & 6 & 1 & 7 \\ 2 & 9 & 1 & 0 \\ 3 & 8 & 1 & 8 \\ 1 & 2 & 1 & 16 \end{bmatrix}$

(iii) $\mathbf{A} = \begin{bmatrix} 6 \\ -2 \end{bmatrix}$ $\mathbf{B} = \begin{bmatrix} 1 & -1 \end{bmatrix}$

3. Find $\mathbf{A}^t$, $\mathbf{A}^t\mathbf{A}$, $\mathbf{AA}^t$:

(i) $\mathbf{A} = \begin{bmatrix} 2 & -1 & 3 \\ 4 & 3 & -5 \end{bmatrix}$

(ii) $\mathbf{A} = \begin{bmatrix} 6 & 0 \\ 0 & -4 \\ 7 & 5 \end{bmatrix}$

4. Find the inverse, if it exists, of the matrices

(i) $\begin{bmatrix} 2 & 3 \\ 1 & 4 \end{bmatrix}$

(ii) $\begin{bmatrix} 1 & 2 & 0 \\ 0 & 1 & 0 \\ 0 & 3 & 2 \end{bmatrix}$

5. Find the determinants of the following matrices.

(i) $\begin{bmatrix} 1 & 0 & 1 & 0 \\ 3 & 2 & -7 & 1 \\ 6 & 1 & -6 & 1 \\ 2 & 2 & 2 & 3 \end{bmatrix}$

(ii) $\begin{bmatrix} 4 & 1 & 6 & 2 \\ 0 & 1 & 4 & 2 \\ 0 & 0 & 9 & 0 \\ 0 & 0 & 0 & -1 \end{bmatrix}$

(iii) $\begin{bmatrix} 3 & 0 & 4 \\ 1 & 0 & 0 \\ 0 & 0 & 6 \end{bmatrix}$

6. Solve the system of equations

$$\begin{aligned} x + 2y - z &= 2 \\ 8x + 3y - 7z &= 4 \\ 4y - 12z &= -8. \end{aligned}$$

7. First showing that a non-trivial solution does indeed exist, solve

$$\begin{aligned} 4x - y &= 0 \\ 4y - z &= 0 \\ -4x + 17y - 4z &= 0. \end{aligned}$$

8. For what values of a and c do you get

(i) one solution,

(ii) no solution,

(iii) infinite solutions,

for the system

$$\begin{aligned} x + 5y + z &= 0 \\ x + 6y - z &= 2 \\ 2x + ay + z &= c\ ? \end{aligned}$$

9. Find the eigenvalues and eigenvectors of $\mathbf{A}$:

(i)

$$\mathbf{A} = \begin{bmatrix} 2 & 0 & 1 \\ 0 & 3 & 4 \\ 0 & 0 & 1 \end{bmatrix}$$

(ii)

$$\mathbf{A} = \begin{bmatrix} 2 & 1 & 0 \\ 1 & 2 & 0 \\ 0 & 0 & 1 \end{bmatrix}$$

CHAPTER 7
VECTORS

7.1 VECTOR NOTATION

A vector in R^2 is represented by an ordered pair $\underset{\sim}{v} = (a, b)$, or geometrically, by a directed line segment in the plane.

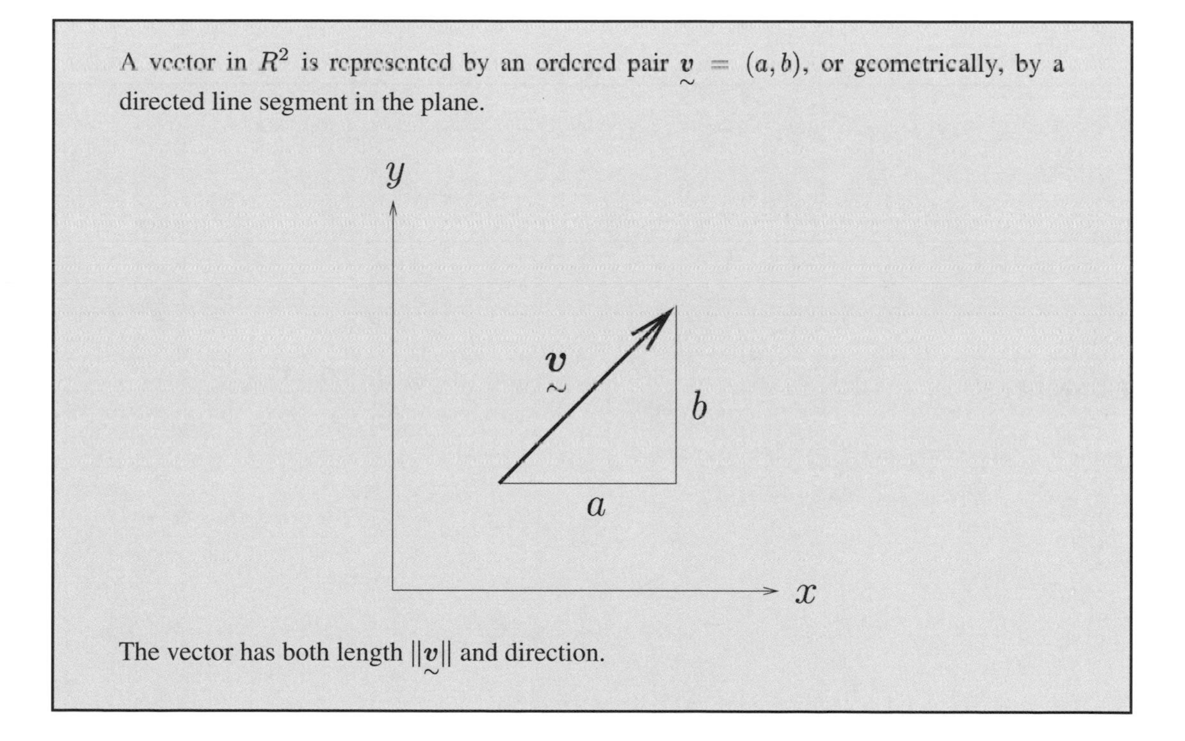

The vector has both length $||\underset{\sim}{v}||$ and direction.

EXAMPLES

1. The vector $(1, 0)$ points in the x direction and has length 1.

2. The vector $(1, 1)$ points in a direction with angle $\pi/4$ to the x axis.

A vector in R^n is represented by an ordered n-tuple

$$\underset{\sim}{\boldsymbol{v}} = (v_1, v_2, \ldots, v_n).$$

EXAMPLES

1. $\underset{\sim}{\boldsymbol{v}} = (1, 3, 4)$ is a three dimensional vector so $\underset{\sim}{\boldsymbol{v}} \in R^3$.

2. $\underset{\sim}{\boldsymbol{v}} = (3, 5, 7, 1, 2)$ is a five dimensional vector so $\underset{\sim}{\boldsymbol{v}} \in R^5$.

7.2 ADDITION AND SCALAR MULTIPLICATION

If $\underset{\sim}{\boldsymbol{v}} = (v_1, v_2, \ldots, v_n)$, $\underset{\sim}{\boldsymbol{w}} = (w_1, w_2, \ldots, w_n)$ and c is a scalar constant then

$$\begin{aligned} \underset{\sim}{\boldsymbol{v}} + \underset{\sim}{\boldsymbol{w}} &= (v_1 + w_1, v_2 + w_2, \ldots, v_n + w_n) \\ c\underset{\sim}{\boldsymbol{v}} &= (cv_1, cv_2, \ldots, cv_n). \end{aligned}$$

EXAMPLES

1. If $\underset{\sim}{\boldsymbol{v}} = (1, -1, 4)$ and $\underset{\sim}{\boldsymbol{w}} = (1, -3, 3)$ then,

$$\underset{\sim}{\boldsymbol{v}} + \underset{\sim}{\boldsymbol{w}} = (2, -4, 7),$$

and

$$4\underset{\sim}{\boldsymbol{v}} = (4, -4, 16).$$

2. If $\underset{\sim}{\boldsymbol{i}} = (1, 0)$ and $\underset{\sim}{\boldsymbol{j}} = (0, 1)$ then $\underset{\sim}{\boldsymbol{v}} = (2, 3) = 2\underset{\sim}{\boldsymbol{i}} + 3\underset{\sim}{\boldsymbol{j}}$.

3. If $\underset{\sim}{\boldsymbol{v}} = (1, x, 2, x)$ and $\underset{\sim}{\boldsymbol{u}} = (2, 1, x, 0)$ then $\underset{\sim}{\boldsymbol{v}} + \underset{\sim}{\boldsymbol{u}} = (3, 1 + x, 2 + x, x)$.

4. $(1, 2) + (4, 5, 6)$ is not defined.

7.3 LENGTH

The *length* of a vector in R^n is given by

$$\|\underset{\sim}{v}\| = \sqrt{v_1^2 + v_2^2 + \cdots + v_n^2}\,.$$

EXAMPLES

1. $\|(1,2,1)\| = \sqrt{1^2 + 2^2 + 1^2} = \sqrt{6}$
2. $\|(1,1,2,1,3)\| = \sqrt{1+1+4+1+9} = \sqrt{16} = 4$

The **triangle inequality** states that

$$\|\underset{\sim}{u} + \underset{\sim}{v}\| \leq \|\underset{\sim}{u}\| + \|\underset{\sim}{v}\|.$$

That is the length of the sum of vectors must be less than the length of the two individual vectors added.

EXAMPLE

$(0,3) + (4,0) = (4,3)$ and $\|(0,3)\| = 3$, $\|(4,0)\| = 4, \|(4,3)\| = \sqrt{3^2 + 4^2} = 5 < 3 + 4$.

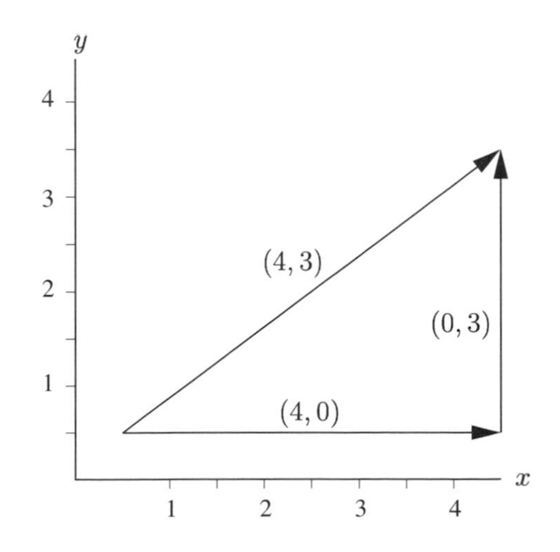

7.4 CARTESIAN UNIT VECTORS

The Cartesian unit vectors for R^3 are

$$\underset{\sim}{\boldsymbol{i}} = (1,0,0), \quad \underset{\sim}{\boldsymbol{j}} = (0,1,0), \quad \underset{\sim}{\boldsymbol{k}} = (0,0,1).$$

Vectors in R^3 are often written as the sum of the components in the direction of the Cartesian unit vectors:

$$\underset{\sim}{\boldsymbol{v}} = (v_1, v_2, v_3) = v_1 \underset{\sim}{\boldsymbol{i}} + v_2 \underset{\sim}{\boldsymbol{j}} + v_3 \underset{\sim}{\boldsymbol{k}} \,.$$

EXAMPLES

1. $(1,2,3) = \underset{\sim}{\boldsymbol{i}} + 2\underset{\sim}{\boldsymbol{j}} + 3\underset{\sim}{\boldsymbol{k}}$

2. $(0,2,0) = 2\underset{\sim}{\boldsymbol{j}}$

7.5 DOT PRODUCT

If $\underset{\sim}{\boldsymbol{u}}$ and $\underset{\sim}{\boldsymbol{v}}$ are vectors in R^n then the *dot product* is defined by

$$\underset{\sim}{\boldsymbol{u}} \cdot \underset{\sim}{\boldsymbol{v}} = u_1 v_1 + u_2 v_2 + \cdots + u_n v_n \,.$$

This is also called an *inner product* on R^n. The result of a dot product is a *scalar*.

EXAMPLES

1. $(1,2,3) \cdot (1,1,1) = 1 + 2 + 3 = 6$

2. $(1,2,3) \cdot (1,2,3) = 1^2 + 2^2 + 3^2 = 14$

3. $||\underset{\sim}{\boldsymbol{u}}||^2 = \underset{\sim}{\boldsymbol{u}} \cdot \underset{\sim}{\boldsymbol{u}}$

The angle θ between two vectors is given by

$$\cos\theta = \frac{\underset{\sim}{\boldsymbol{u}} \cdot \underset{\sim}{\boldsymbol{v}}}{\|\underset{\sim}{\boldsymbol{u}}\|\,\|\underset{\sim}{\boldsymbol{v}}\|}.$$

EXAMPLES

1. The angle θ between $(1,2,3)$ and $(1,1,1)$ is such that

$$\cos\theta = \frac{6}{\sqrt{1^2+2^2+3^2}\,\sqrt{1^2+1^2+1^2}} = \frac{6}{\sqrt{42}}.$$

2. $(1,2,3)$ and $(1,1,-1)$ are at right angles since $(1,2,3)\cdot(1,1,-1)=0$ hence $\cos\theta = 0$.

Two vectors, $\underset{\sim}{\boldsymbol{u}}$ and $\underset{\sim}{\boldsymbol{v}}$ are **orthogonal** if they are perpendicular to each other and

$$\underset{\sim}{\boldsymbol{u}} \cdot \underset{\sim}{\boldsymbol{v}} = 0.$$

EXAMPLES

1. $\underset{\sim}{\boldsymbol{u}} = (1,2,1)$ and $\underset{\sim}{\boldsymbol{v}} = (2,1,-4)$ are perpendicular since $\underset{\sim}{\boldsymbol{u}} \cdot \underset{\sim}{\boldsymbol{v}} = 2+2-4=0$.

2. To find a vector, (a,b), perpendicular to $(1,2)$ write

$$(a,b)\cdot(1,2) = a+2b = 0$$

hence the simplest choice is $(a,b) = (-2,1)$ although any multiple of this will be perpendicular to $(1,2)$. For example $(2,-1)$ and $(-4,2)$ are still perpendicular to $(1,2)$.

7.6 CROSS PRODUCT

If $\underset{\sim}{\boldsymbol{u}}$ and $\underset{\sim}{\boldsymbol{v}}$ are two vectors in R^3, then the *cross product* $\underset{\sim}{\boldsymbol{u}} \times \underset{\sim}{\boldsymbol{v}}$ is defined in determinant notation by

$$\begin{aligned}
\underset{\sim}{\boldsymbol{u}} \times \underset{\sim}{\boldsymbol{v}} &= \begin{vmatrix} \underset{\sim}{\boldsymbol{i}} & \underset{\sim}{\boldsymbol{j}} & \underset{\sim}{\boldsymbol{k}} \\ u_1 & u_2 & u_3 \\ v_1 & v_2 & v_3 \end{vmatrix} \\
&= \underset{\sim}{\boldsymbol{i}} \begin{vmatrix} u_2 & u_3 \\ v_2 & v_3 \end{vmatrix} - \underset{\sim}{\boldsymbol{j}} \begin{vmatrix} u_1 & u_3 \\ v_1 & v_3 \end{vmatrix} + \underset{\sim}{\boldsymbol{k}} \begin{vmatrix} u_1 & u_2 \\ v_1 & v_2 \end{vmatrix} \\
&= (u_2v_3 - u_3v_2,\, u_3v_1 - u_1v_3,\, u_1v_2 - u_2v_1)\,.
\end{aligned}$$

EXAMPLES

1. $(1,0,0) \times (0,1,0) = (0,0,1)$. That is $\underset{\sim}{\boldsymbol{i}} \times \underset{\sim}{\boldsymbol{j}} = \underset{\sim}{\boldsymbol{k}}$.

2. If $\underset{\sim}{\boldsymbol{u}} = (2,-3,1)$ and $\underset{\sim}{\boldsymbol{v}} = (12,4,-6)$, then

$$\begin{aligned}
\underset{\sim}{\boldsymbol{w}} = \underset{\sim}{\boldsymbol{u}} \times \underset{\sim}{\boldsymbol{v}} &= \begin{vmatrix} \underset{\sim}{\boldsymbol{i}} & \underset{\sim}{\boldsymbol{j}} & \underset{\sim}{\boldsymbol{k}} \\ 2 & -3 & 1 \\ 12 & 4 & -6 \end{vmatrix} \\
&= \underset{\sim}{\boldsymbol{i}} \begin{vmatrix} -3 & 1 \\ 4 & -6 \end{vmatrix} - \underset{\sim}{\boldsymbol{j}} \begin{vmatrix} 2 & 1 \\ 12 & -6 \end{vmatrix} + \underset{\sim}{\boldsymbol{k}} \begin{vmatrix} 2 & -3 \\ 12 & 4 \end{vmatrix} \\
&= \underset{\sim}{\boldsymbol{i}}(18-4) - \underset{\sim}{\boldsymbol{j}}(-12-12) + \underset{\sim}{\boldsymbol{k}}(8+36) \\
&= 14\underset{\sim}{\boldsymbol{i}} + 24\underset{\sim}{\boldsymbol{j}} + 44\underset{\sim}{\boldsymbol{k}} = (14,24,44).
\end{aligned}$$

3. The cross product $(1,2,3) \times (1,0,1)$ is:

$$(1,2,3) \times (1,0,1) = \begin{vmatrix} \underset{\sim}{\boldsymbol{i}} & \underset{\sim}{\boldsymbol{j}} & \underset{\sim}{\boldsymbol{k}} \\ 1 & 2 & 3 \\ 1 & 0 & 1 \end{vmatrix} = (2,2,-2).$$

4. The cross product $(1,1,2) \times (1,-1,0)$ is:

$$(1,1,2) \times (1,-1,0) = \begin{vmatrix} \underset{\sim}{\boldsymbol{i}} & \underset{\sim}{\boldsymbol{j}} & \underset{\sim}{\boldsymbol{k}} \\ 1 & 1 & 2 \\ 1 & -1 & 0 \end{vmatrix} = (2,2,-2).$$

> Note that the result of taking the cross product of two vectors is another *vector* where the *direction* of $\underset{\sim}{u} \times \underset{\sim}{v}$ is *perpendicular* to both $\underset{\sim}{u}$ and $\underset{\sim}{v}$.

EXAMPLES

1. In a previous example $(2,-3,1) \times (12,4,-6) = (14,24,44)$ and

$$(2,3,-1) \cdot (14,24,44) = 28 - 72 + 44 = 0.$$

   Similarly $(12,4,-6) \cdot (14,24,44) = 0$.

2. In a previous example $(1,1,2) \times (1,-1,0) = (2,2,-2)$. Note that $(2,2,-2) \cdot (1,1,2) = 0$ and $(2,2,-2) \cdot (1,-1,0) = 0$.

7.7 LINEAR INDEPENDENCE

> A vector $\underset{\sim}{v}$ is a **linear combination** of the vectors $\underset{\sim}{u_1}, \underset{\sim}{u_2}, \ldots, \underset{\sim}{u_n}$ if it can be written as
>
> $$\underset{\sim}{v} = c_1\underset{\sim}{u_1} + c_2\underset{\sim}{u_2} + \cdots + c_n\underset{\sim}{u_n}$$
>
> where $c_1, \ldots, c_n$ are constants.

EXAMPLES

1. $(2,7,3)$ is a linear combination of $(1,1,0), (0,2,1), (0,1,0)$ since

$$(2,7,3) = 2(1,1,0) + 3(0,2,1) - (0,1,0).$$

2. $(1,2,1)$ is not a linear combination of $(1,1,0), (2,1,0), (1,0,0)$ since we can never combine the three vectors to get the third component of $(1,2,1)$.

3. Any vector (a,b,c) in R^3 can be found from a linear combination of $\{(1,0,0), (0,1,0), (0,0,1)\}$.

A set of vectors $\underset{\sim}{u_1}, \underset{\sim}{u_2}, \ldots, \underset{\sim}{u_n}$ are **linearly independent** if the only constants $c_1, \ldots, c_n$ that satisfy

$$c_1\underset{\sim}{u_1} + c_2\underset{\sim}{u_2} + \cdots + c_n\underset{\sim}{u_n} = \underset{\sim}{\mathbf{0}}$$

are $c_1 = c_2 = \cdots = c_n = 0$.

EXAMPLES

1. $(1, 1, 0), (0, 2, 1), (0, 1, 0)$ are independent since

$$c_1(1, 1, 0) + c_2(0, 2, 1) + c_3(0, 1, 0) = (0, 0, 0)$$

implies

$$\begin{aligned} c_1 &= 0 \\ c_1 + c_2 + c_3 &= 0 \\ c_2 &= 0 \end{aligned}$$

which gives $c_1 = c_2 = c_3 = 0$.

2. $(1, 1, 0), (2, 1, 0), (1, 0, 0)$ are dependent (*not* linearly independent) since

$$\begin{aligned} c_1 + 2c_2 + c_3 &= 0 \\ c_1 + c_2 &= 0 \\ 0 &= 0 \end{aligned}$$

has an infinite number of solutions, one being $c_1 = 1,\ c_2 = -1,\ c_3 = 1$.

3. The vectors $(1, 2), (2, 1), (1, 0)$ are dependent since

$$c_1(1, 2) + c_2(2, 1) + c_3(1, 0) = (0, 0)$$

implies

$$\begin{aligned} c_1 + 2c_2 + c_3 &= 0 \\ 2c_1 + 2c_2 &= 0. \end{aligned}$$

Since we have two equations in three unknowns we can always find a non-zero c_1, c_2, c_3 to satisfy these equations, for example $c_1 = 1,\ c_2 = -1,\ c_3 = 1$. More than two vectors in R^2 can never be independent.

4. The vectors $\underset{\sim}{i}, \underset{\sim}{j}, \underset{\sim}{k}$ are independent since for any vector $\underset{\sim}{v} = (a, b, c)$ it is possible to write

$$(a, b, c) = c_1\underset{\sim}{i} + c_2\underset{\sim}{j} + c_3\underset{\sim}{k} = a\underset{\sim}{i} + b\underset{\sim}{j} + c\underset{\sim}{k}$$

hence if $\underset{\sim}{v} = \underset{\sim}{\mathbf{0}}$ then $c_1 = c_2 = c_3 = 0$.

A set of vectors is linearly independent if the determinant of the matrix with vectors as columns is not zero.

EXAMPLES

1. For $(1,1,0), (0,2,1), (0,1,0)$ the determinant

$$\begin{vmatrix} 1 & 0 & 0 \\ 1 & 2 & 1 \\ 0 & 1 & 0 \end{vmatrix} = -1 \neq 0$$

hence the vectors are independent.

2. For $(1,1,0), (2,1,0), (1,0,0)$ the determinant

$$\begin{vmatrix} 1 & 2 & 1 \\ 1 & 1 & 0 \\ 0 & 0 & 0 \end{vmatrix} = 0$$

hence the vectors are dependent. We can show that

$$(2,1,0) = (1,1,0) + (1,0,0)$$

so they are not independent of each other.

7.8 EXAMPLE QUESTIONS

(Answers are given in Chapter 14)

1. Evaluate the sum $\underset{\sim}{\boldsymbol{u}} + \underset{\sim}{\boldsymbol{v}}$, $3\underset{\sim}{\boldsymbol{u}}$ and $||\underset{\sim}{\boldsymbol{u}}||$:

   (i) $\underset{\sim}{\boldsymbol{u}} = (-2, -1)$, $\underset{\sim}{\boldsymbol{v}} = (1, 1)$

   (ii) $\underset{\sim}{\boldsymbol{u}} = (3, 4)$, $\underset{\sim}{\boldsymbol{v}} = (4, 3)$

   (iii) $\underset{\sim}{\boldsymbol{u}} = (-2, 1)$, $\underset{\sim}{\boldsymbol{v}} = (-1, -1)$

   (iv) $\underset{\sim}{\boldsymbol{u}} = (3, 4, 2)$, $\underset{\sim}{\boldsymbol{v}} = (1, 1, 1)$

   (v) $\underset{\sim}{\boldsymbol{u}} = (3, 1, 1, 0)$, $\underset{\sim}{\boldsymbol{v}} = (1, 0, 1, 1)$

   (vi) $\underset{\sim}{\boldsymbol{u}} = 2\underset{\sim}{\boldsymbol{i}} + 3\underset{\sim}{\boldsymbol{j}} + \underset{\sim}{\boldsymbol{k}}$, $\underset{\sim}{\boldsymbol{v}} = \underset{\sim}{\boldsymbol{i}} - \underset{\sim}{\boldsymbol{j}} - \underset{\sim}{\boldsymbol{k}}$

   (vii) $\underset{\sim}{\boldsymbol{u}} = \underset{\sim}{\boldsymbol{i}} + \underset{\sim}{\boldsymbol{j}}$, $\underset{\sim}{\boldsymbol{v}} = \underset{\sim}{\boldsymbol{i}} - 3\underset{\sim}{\boldsymbol{j}}$

2. For the above vectors verify the triangle inequality that $||\underset{\sim}{\boldsymbol{u}} + \underset{\sim}{\boldsymbol{v}}|| \leq ||\underset{\sim}{\boldsymbol{u}}|| + ||\underset{\sim}{\boldsymbol{v}}||$.

3. In the diagram below write down the two vectors $\underset{\sim}{\boldsymbol{u}}$ and $\underset{\sim}{\boldsymbol{v}}$ in algebraic form then find and draw the vector $\underset{\sim}{\boldsymbol{u}} + \underset{\sim}{\boldsymbol{v}}$.

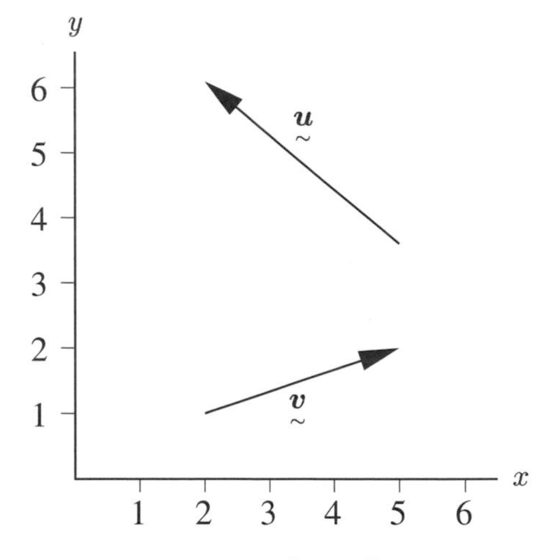

4. Evaluate the sum $\underset{\sim}{\boldsymbol{u}} + \underset{\sim}{\boldsymbol{v}}$ and $||\underset{\sim}{\boldsymbol{u}} + \underset{\sim}{\boldsymbol{v}}||$:

   (i) $\underset{\sim}{\boldsymbol{u}} = (3, 2, -1)$, $\underset{\sim}{\boldsymbol{v}} = (-1, -2, 1)$

   (ii) $\underset{\sim}{\boldsymbol{u}} = (1, 0, 9)$, $\underset{\sim}{\boldsymbol{v}} = (-2, -2, -2)$

   (iii) $\underset{\sim}{\boldsymbol{u}} = (4, -4, -3)$, $\underset{\sim}{\boldsymbol{v}} = (8, 7, 1)$

5. Find $\underset{\sim}{\boldsymbol{u}} \cdot \underset{\sim}{\boldsymbol{v}}$, $\underset{\sim}{\boldsymbol{u}} \times \underset{\sim}{\boldsymbol{v}}$ and $\cos\theta$ where θ is the angle between the $\underset{\sim}{\boldsymbol{u}}$ and $\underset{\sim}{\boldsymbol{v}}$:

   (i) $\underset{\sim}{\boldsymbol{u}} = (1, 2, 1)$, $\underset{\sim}{\boldsymbol{v}} = (-1, 3, 1)$

   (ii) $\underset{\sim}{\boldsymbol{u}} = (-3, 2, -1)$, $\underset{\sim}{\boldsymbol{v}} = (6, 1, 1)$

   (iii) $\underset{\sim}{\boldsymbol{u}} = (2, 3, 0)$, $\underset{\sim}{\boldsymbol{v}} = (4, 1, -2)$

   (iv) $\underset{\sim}{\boldsymbol{u}} = (0, 0, 0)$, $\underset{\sim}{\boldsymbol{v}} = (1, 4, 3)$

   (v) $\underset{\sim}{\boldsymbol{u}} = (3, 3, 3)$, $\underset{\sim}{\boldsymbol{v}} = (-1, -1, -1)$

   (vi) $\underset{\sim}{\boldsymbol{u}} = (1, 2, 4)$, $\underset{\sim}{\boldsymbol{v}} = (2, 4, -2)$

6. For the previous question verify that $\underset{\sim}{\boldsymbol{w}} = \underset{\sim}{\boldsymbol{u}} \times \underset{\sim}{\boldsymbol{v}}$ is orthogonal (at right angles) to both $\underset{\sim}{\boldsymbol{u}}$ and $\underset{\sim}{\boldsymbol{v}}$.

7. Determine whether the following vectors are linearly independent

   (i) $\{(4, 1), (1, 2)\}$

   (ii) $\{(2, 1), (4, 2)\}$

   (iii) $\{(1, 1), (1, 2), (3, 1)\}$

   (iv) $\{(1, 1, 1), (0, 2, 0), (1, 3, 2)\}$

   (v) $\{(1, 1, 1), (0, 2, 0), (1, 3, 1)\}$

   (vi) $\{(1, 2, 0, 1), (1, 1, 0, 1), (2, 1, 3, 1), (0, 2, -3, 1)\}$

8. Find a number c so that $(1, 2, c)$ is orthogonal to $(2, 1, 2)$.

9. Find the vector which goes from the *point* $(1, 3, 1)$ to the *point* $(2, 5, 3)$. What is the length of this vector?

10. Show that the line through the *points* $(1, 1, 1)$ and $(2, 3, 4)$ is perpendicular to the line through the *points* $(1, 0, 0)$ and $(3, -1, 0)$.

11. Show that $\mathbf{a} \cdot (\mathbf{b} \times \mathbf{c})$ can be written as

$$\mathbf{a} \cdot (\mathbf{b} \times \mathbf{c}) = \begin{vmatrix} a_1 & a_2 & a_3 \\ b_1 & b_2 & b_3 \\ c_1 & c_2 & c_3 \end{vmatrix}$$
$$= a_1 b_2 c_3 - a_1 b_3 c_2 - a_2 b_1 c_3 + a_2 b_3 c_1 + a_3 b_1 c_2 - a_3 b_2 c_1.$$

12. Verify the above equation using the vectors $\mathbf{a} = (1, 1, 2)$, $\mathbf{b} = (1, 0, 1)$, $\mathbf{c} = (0, 1, 1)$.

CHAPTER 8

ASYMPTOTICS AND APPROXIMATIONS

8.1 LIMITS

As $x \to 0$ then

1. $x^n < x^m \quad$ if $1 < m < n, \quad 0 < x < 1$
2. $\lim_{x \to 0} f(x) + g(x) = \lim_{x \to 0} f(x) + \lim_{x \to 0} g(x)$
3. $\lim_{x \to 0} f(x)g(x) = \lim_{x \to 0} f(x) \lim_{x \to 0} g(x)$

assuming $\lim_{x \to 0} f(x)$ and $\lim_{x \to 0} g(x)$ exist.

EXAMPLES

1. $(0.1)^3 < (0.1)^2$

2. $x^3 < x^2$ as $x \to 0$.

3. $\lim_{x \to 0} \sin x \cos x = \lim_{x \to 0} \sin x \ \lim_{x \to 0} \cos x = 0 \times 1 = 0$

4. $\lim_{x \to 0} \dfrac{x(x-1)}{x(x-2)} = \dfrac{1}{2}$

8.2 L'HÔPITAL'S RULE

If $\dfrac{f(x)}{g(x)}$ has limit $\dfrac{0}{0}$ or $\dfrac{\infty}{\infty}$ as $x \to x_c$ then

$$\lim_{x \to x_c} \frac{f(x)}{g(x)} = \lim_{x \to x_c} f'(x) / \lim_{x \to x_c} g'(x).$$

EXAMPLES

1.
$$\lim_{x \to 1} \frac{x^2 - 1}{x - 1} = \lim_{x \to 1} \frac{2x}{1} = 2$$

2.
$$\lim_{x \to 0} \frac{\sin x}{x} = \lim_{x \to 0} \frac{\cos x}{1} = 1$$

8.3 TAYLOR SERIES

$$\begin{aligned} f(x) &= f(0) + x f'(0) + \frac{x^2}{2!} f''(x) + \cdots \quad \text{Maclaurin series} \\ f(x+a) &= f(a) + (x-a) f'(a) + \frac{(x-a)^2}{2!} f''(a) + \cdots \quad \text{Taylor series} \end{aligned}$$

EXAMPLES

1.
$$\begin{aligned} \sin x &= \sin 0 + x \sin' 0 + \frac{x^2}{2} \sin'' 0 + \frac{x^3}{6} \sin''' 0 + \cdots \\ &= \sin 0 + x \cos 0 - \frac{x^2}{2} \sin 0 - \frac{x^3}{6} \cos 0 + \cdots \\ &= x - \frac{x^3}{6} + \cdots \end{aligned}$$

2.
$$e^x = 1 + x + \frac{x^2}{2} + \frac{x^3}{6} + \cdots$$

8.4 ASYMPTOTICS

As $x \to +\infty$ then

$$\begin{aligned} x^m &< x^n, && \text{if } m < n \\ x^m &< e^{ax}, && \text{if } a > 0 \\ x^m &> e^{ax}, && \text{if } a < 0. \end{aligned}$$

EXAMPLES

1.
$$(100)^2 < (100)^3$$

2.
$$(100)^{0.2} < (100)^{0.5}$$

3.
$$x^0 < e^{x/10} \quad \text{as} \quad x \to \infty.$$

4.
$$\frac{1}{x^2} > e^{-x} \quad \text{as} \quad x \to \infty.$$

5.
$$\frac{x^2+1}{2x^2+x+3} \to \frac{x^2}{2x^2} \to \frac{1}{2} \quad \text{as} \quad x \to \infty.$$

6.
$$\frac{x^2+1}{2x^3+x+3} \sim \frac{x^2}{2x^3} \sim \frac{1}{2x} \quad \text{as} \quad x \to \infty.$$

7.
$$\frac{xe^x}{1+e^{2x}} \sim \frac{xe^x}{e^{2x}} \sim xe^{-x} \quad \text{as} \quad x \to \infty.$$

8.
$$\frac{e^{2x}}{\cosh x} \sim \frac{2e^{2x}}{e^x} \sim 2e^x \quad \text{as} \quad x \to \infty.$$

9.
$$\frac{xe^x}{x+\sqrt{x^2+2x+1}} \sim \frac{xe^x}{x+\sqrt{x^2}} \sim \frac{e^x}{2} \quad \text{as} \quad x \to \infty.$$

8.5 EXAMPLE QUESTIONS

(Answers are given in Chapter 14)

1. Find the limits as $x \to 0$ of the following functions.

   (i) $(x-2)(x-3)$

   (ii) $x + \cos x$

   (iii) $e^x + xe^{-x}$

   (iv) $\dfrac{\sin(x-\pi)}{(x-\pi)}$

   (v) $\dfrac{x^3+x^2+1}{x^2+3x+2}$

   (vi) $\dfrac{x^2+1}{1+2/x}$

2. Use L'Hôpital's Rule to find the following limits.

   (i) $\displaystyle\lim_{x\to 0} \frac{\sin x - x}{x^3}$

   (ii) $\displaystyle\lim_{x\to 1} \frac{x^3-3x+2}{x^3-1}$

   (iii) $\displaystyle\lim_{x\to 1} \frac{x^3-2x+1}{x^3-1}$

   (iv) $\displaystyle\lim_{x\to 0} \frac{x\cos x - \sin x}{x^3}$

3. Find the first two non-zero terms in the Taylor Series for the following functions as $x \to 0$.

   (i) $\cos x$

   (ii) $\dfrac{\sin x}{x}$

   (iii) $\dfrac{x^2+1}{1+1/x}$

   (iv) xe^x

   (v) $\dfrac{\cos x - 1}{x}$

   (vi) $\sinh x$

   (vii) $\dfrac{\cosh x - 1}{x^2}$

   (viii) $\dfrac{1}{1-x}$

   (ix) $\ln(1+x)$

4. Find the leading order behaviour for the following functions as $x \to \infty$.

   (i) $\dfrac{x^2+x}{3x^2+2x+1}$

   (ii) $\sqrt{x^3+2x+1}$

   (iii) $\dfrac{x^3+2x}{3x^2+\sqrt{x^4+1}}$

   (iv) $\dfrac{x^2+1}{1+1/x}$

   (v) $\dfrac{\sinh(x)}{\cosh(x)}$

   (vi) $\dfrac{\sinh(3x)}{1+e^{4x}}$

   (vii) $\dfrac{xe^{-x}}{\sinh x}$

   (viii) $\dfrac{x+e^{-x}}{1+xe^{-x}}$

   (ix) $\dfrac{e^x}{\sinh x}$

5. Find the first three non-zero terms of the Taylor series as $x \to 0$ for the functions $f(x) = e^x$ and $g(x) = \sin x$ and **hence** find the Maclaurin series for

   (i) $e^x \sin x$

   (ii) e^{2x}

   (iii) $\sin x^2$

   (iv) e^{x^3}

6. Find the first two non-zero terms for the Taylor series about $x = \pi/2$ for

   (i) $\sin x$

   (ii) $\cos x$

7. Find the first three terms in the Maclaurin series for e^{-x^2} and hence find an approximation for

   $$\int_0^{\epsilon} e^{-x^2}\,dx$$

   where ϵ is a small number. (Expand the exponential and then integrate.)

CHAPTER 9

COMPLEX NUMBERS

9.1 DEFINITION

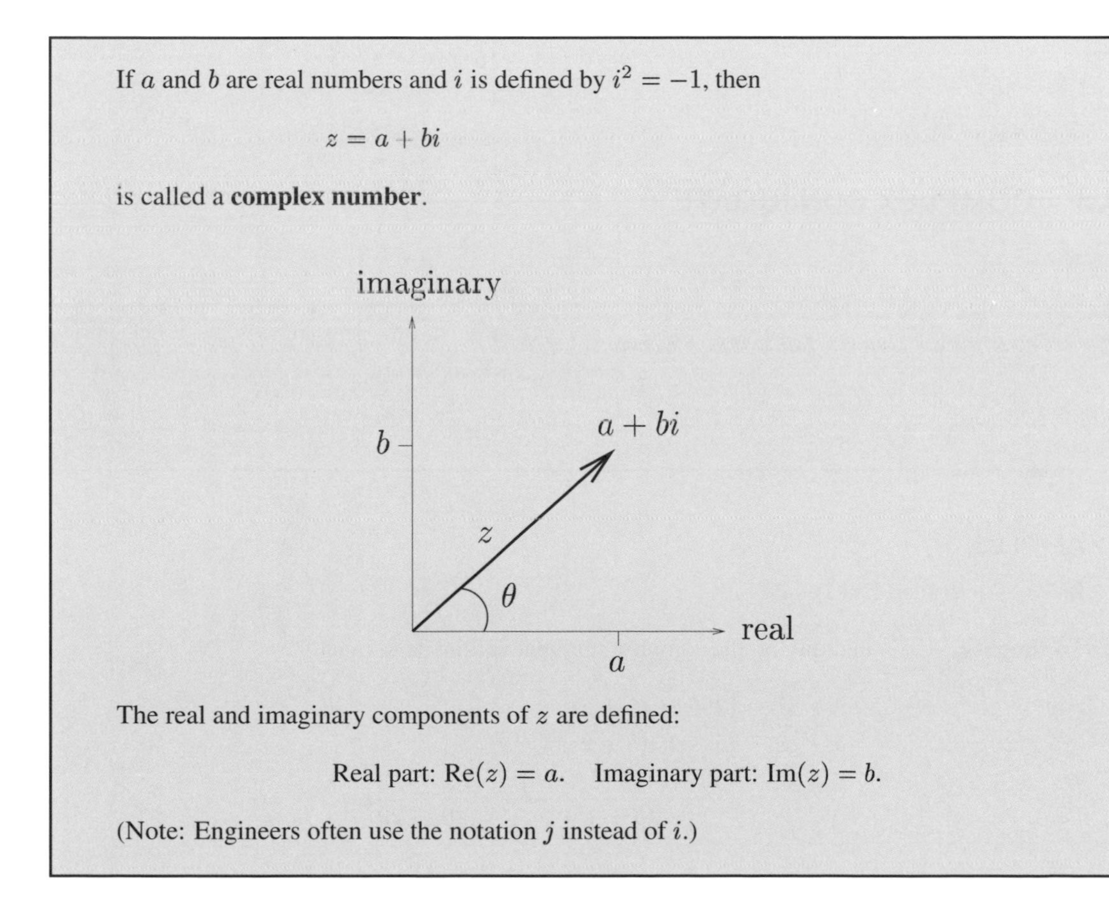

If a and b are real numbers and i is defined by $i^2 = -1$, then

$$z = a + bi$$

is called a **complex number**.

The real and imaginary components of z are defined:

Real part: $\mathrm{Re}(z) = a$. Imaginary part: $\mathrm{Im}(z) = b$.

(Note: Engineers often use the notation j instead of i.)

9.2 ADDITION AND MULTIPLICATION

If $z_1 = a + bi$ and $z_2 = c + di$ then

1. $z_1 \pm z_2 = (a \pm c) + (b \pm d)i$
2. $c(a + bi) = ca + cbi,$ where c is any real number
3. $z_1 z_2 = (a + bi)(c + di) = (ac - bd) + (ad + bc)i.$

EXAMPLES

1. Given $z_1 = 5 - 4i$ and $z_2 = -6 + 2i$ then
$$z_1 + z_2 = (5 - 6) + (-4 + 2)i = -1 - 2i.$$
2. If $z_1 = 5 + 8i$ then $3z_1 = 15 + 24i.$
3. If $z_1 = 5 + 8i$ and $z_2 = -2 + 3i$ then
$$z_1 z_2 = (-10 - 24) + (15 - 16)i = -34 - i.$$

9.3 COMPLEX CONJUGATE

The complex conjugate of $z = a + bi$ is

$$\overline{z} = a - bi.$$

EXAMPLES

1. If $z = 1 + 2i$ then $\overline{z} = 1 - 2i.$
2. To simplify $\dfrac{2 + 3i}{5 - 2i}$ multiply by the complex conjugate of the denominator:
$$\begin{aligned}\frac{2 + 3i}{5 - 2i} &= \frac{(2 + 3i)(5 + 2i)}{(5 - 2i)(5 + 2i)}\\ &= \frac{(10 - 6) + (4 + 15)i}{25 + 4} = \frac{4}{29} + \frac{19}{29}i.\end{aligned}$$

9.4 EULER'S EQUATION

In Polar form a complex number is written as:

$$z = r(\cos\theta + i\sin\theta) = re^{i\theta} \quad (\equiv r\,\mathrm{cis}\,\theta),$$

$$\text{where} \qquad r = |z| = \sqrt{a^2+b^2} \qquad \text{and} \qquad \theta = \tan^{-1}\left(\frac{b}{a}\right).$$

Using this notation,

$$z_1 z_2 = r_1 r_2 e^{i(\theta_1+\theta_2)}.$$

Note that $\theta \in [0, 2\pi)$ can be replaced by $\theta + 2k\pi, k = 0, 1, 2\ldots$

EXAMPLES

1. $(1+i) = \sqrt{2}e^{i\pi/4} \equiv \sqrt{2}\,\mathrm{cis}\,\frac{\pi}{4}$
2. $-1-i = \sqrt{2}e^{i5\pi/4} \equiv \sqrt{2}\,\mathrm{cis}\,\frac{5\pi}{4}$
3. $3e^{i\pi/2} = 3\left(\cos\frac{\pi}{2} + i\sin\frac{\pi}{2}\right) = 3i$
4. $e^{i\pi/4}e^{i3\pi/4} = e^{i\pi} = -1$
5. $2\,\mathrm{cis}\,\frac{\pi}{6} = 2\cos\frac{\pi}{6} + 2i\sin\frac{\pi}{6} = \sqrt{3}+i$
6. If $z = 2+2i$, then

$$r = |z| = \sqrt{2^2+2^2} = \sqrt{8} = 2\sqrt{2}$$
$$\theta = \tan^{-1}\frac{2}{2} = \tan^{-1} 1 = \frac{\pi}{4}$$

so that

$$z = 2+2i = 2\sqrt{2}\exp\left(i\frac{\pi}{4}\right).$$

7. If $z_1 = 2\,\mathrm{cis}\,\frac{\pi}{2}$ and $z_2 = 3\,\mathrm{cis}\,\frac{3\pi}{4}$,

$$\begin{aligned} z_1 z_2 &= 6\exp i\left(\frac{\pi}{2} + \frac{3\pi}{4}\right) \\ &= 6\exp\left(i\frac{5\pi}{4}\right) \\ &\equiv -3\sqrt{2}(1+i). \end{aligned}$$

9.5 DE MOIVRE'S THEOREM

De Moivre's theorem states that

$$z^n = r^n e^{in\theta} \quad (\equiv r^n \operatorname{cis} n\theta).$$

EXAMPLES

1. $(2i)^3 = (2e^{i\pi/2})^3 = 2^3 e^{3i\pi/2} = -8i.$
2. If $z = 1 + i \equiv \sqrt{2}e^{i\pi/4}$ then $z^2 = 2e^{i\pi/2} \equiv 2i.$
3. To find all the cubic roots of -1 first write

$$z = -1 = e^{i\pi} \equiv e^{i(\pi+2k\pi)}$$

where $k = 0, \pm 1, \pm 2, \ldots$ Thus

$$\begin{aligned} z^{1/3} &= \exp\left(i\frac{\pi}{3} + i\frac{2k\pi}{3}\right) \\ &= \exp\left(i\frac{\pi}{3}\right), \exp\left(-i\frac{\pi}{3}\right), \exp\left(i\pi\right) \end{aligned}$$

for $k = 0, 1, -1$. All other values of k will repeat these three solutions. Thus, the 3 cubic roots are

$$\begin{aligned} z_1 &= e^{-i\pi/3}, \quad z_2 = e^{i\pi/3}, \quad z_3 = e^{i\pi} \\ &= \frac{1-\sqrt{3}i}{2}, \quad \frac{1+\sqrt{3}i}{2}, \quad -1. \end{aligned}$$

4. To find z such that $z^2 = i$ we write

$$\begin{aligned} & z^2 = i = e^{i\pi/2+2k\pi}, \quad k = 0, 1, 2, \ldots \\ \Longrightarrow \quad & z = e^{i\pi/4+k\pi} \\ & \quad = e^{i\pi/4}, \quad e^{5i\pi/4} \\ & \quad = \frac{1}{\sqrt{2}}(1+i), \quad \frac{1}{\sqrt{2}}(-1-i). \end{aligned}$$

9.6 EXAMPLE QUESTIONS

(Answers are given in Chapter 14)

1. Write the following in the form $a + bi$.

   (i) $(2 - 3i) + (7 - 9i) - (3 + 13i)$

   (ii) $(1 + i) + (5 - i) + (6 + 6i)$

   (iii) $5(8 - 9i) + i(12 - 42i)$

   (iv) $3(2 - i) + 2i(1 - i)$

   (v) $(7 + 6i)^2$

   (vi) $(1 - i)^3$

   (vii) $(8 - 2i)(3 + 5i)$

   (viii) $(1 + i)(1 - i)$

   (ix) $(2 + 3i)(2 - 3i)$

   (x) $(1 - 4i)(2 + i)$

   (xi) $\dfrac{6 - 2i}{2 - 7i}$

   (xii) $\dfrac{3 - 3i}{3 + 3i}$

   (xiii) $\dfrac{2 - i}{3 + i}$

   (xiv) $\dfrac{3 - i}{(2 - i)(5 + 2i)}$

2. Find x, y if

   (i) $2x + 3yi = (7 - i)(1 + i)$

   (ii) $(x^2 + 6) + xi = 3(2 + 3i)(1 - i)$

3. Write the following in polar form.

   (i) $z = 1$

   (ii) $z = -i$

   (iii) $z = 3i$

   (iv) $z = -1 + i$

   (v) $z = 1 + i$

   (vi) $z = 1 - i$

   (vii) $z = -2 - 2i$

   (viii) $z = \frac{5}{2}(\sqrt{3} - i)$

   (ix) $z = 1 + \sqrt{3}i$

   (x) $-\sqrt{3} + 3i$

   (xi) $z = 5 + 5i$

   (xii) $z = \sqrt{3} + i$

   (xiii) $z = -\sqrt{6} - i\sqrt{2}$

4. Write the following in the form $z = x + iy$.

   (i) $2e^{i\pi/2}$

   (ii) $3e^{i\pi/4}$

   (iii) $\text{cis}\,\dfrac{\pi}{6}$

   (iv) $\text{cis}\,\dfrac{11\pi}{6}$

   (v) $e^{i\pi/3}$

5. Find $z_1 z_2$ and z_1^2 if

   (i) $z_1 = e^{i\pi/2}, z_2 = e^{i\pi/4}$

   (ii) $z_1 = e^{i\pi/6}, z_2 = e^{i\pi/3}$

   (iii) $z_1 = 2e^{i\pi/4}\ z_2 = 3e^{i3\pi/4}$

   (iv) $z_1 = \text{cis}\,\dfrac{\pi}{6}\ z_2 = 3\,\text{cis}\,\dfrac{-5\pi}{6}$

6. Use De Moivre's theorem to find all z where

   (i) $z^2 = i$

   (ii) $z^4 = 1$

   (iii) $z^3 = 4\sqrt{2}(-1 + i)$

   (iv) $z^2 = -i$

   (v) $z^6 = -1$

7. Use Euler's theorem to show

   (i) $\cosh ix = \cos x$

   (ii) $\sinh ix = i \sin x$

   (iii) $\cos(x + y) = \cos x \cos y - \sin x \sin y$

   (iv) $\sin(x + y) = \sin x \cos y + \cos x \sin y$

8. Use De Moivre's theorem to show

   (i) $z^n + z^{-n} = 2\cos n\theta$

   (ii) $\cos 2x = \cos^2 x - \sin^2 x$

   (iii) $\sin 2x = 2 \sin x \cos x$

9. Show that the solution to

   $$az^2 + bz + c = 0,$$

   where a, b, c are all real, must be of the form

   $$z = x + iy, \quad z = x - iy.$$

10. Show that

    $$i(2 + 2i)^6 = 2^9.$$

CHAPTER 10

DIFFERENTIAL EQUATIONS

10.1 FIRST ORDER DIFFERENTIAL EQUATIONS

10.1.1 INTEGRABLE

If

$$\frac{dy}{dx} = g(x)$$

then

$$y = \int g(x)\,dx + c.$$

EXAMPLES

1. To solve $\frac{dy}{dx} - \sin x + x^2 = 0$ for $y(x)$ write

$$\begin{aligned} \frac{dy}{dx} &= \sin x - x^2 \\ \implies \quad y(x) &= \int (\sin x - x^2)\,dx \\ &= -\cos x - \frac{1}{3}x^3 + c. \end{aligned}$$

10.1.2 SEPARABLE

If

$$\frac{dy}{dx} = \frac{g(x)}{h(y)}$$

then

$$\int h(y)\,dy = \int g(x)\,dx.$$

Remember the integrating constant, $+c$, when integrating.

EXAMPLES

1. To solve $\dfrac{dy}{dx} = \dfrac{y}{1+x}$ write

$$\begin{aligned}\int \frac{1}{y}\,dy &= \int \frac{1}{1+x}\,dx\\ \implies \ln|y| &= \ln|1+x| + c\\ \implies y &= (1+x)e^c\\ &= k(1+x), \qquad \text{where } k = e^c \text{ is a constant.}\end{aligned}$$

2. To solve $\dfrac{dy}{dx} = \dfrac{-x}{y}$ with the condition that $y(1) = 1$, write

$$\begin{aligned}\int y\,dy &= -\int x\,dx\\ \implies \frac{1}{2}y^2 &= -\frac{1}{2}x^2 + c\end{aligned}$$

yielding the family of solutions,

$$y^2 + x^2 = 2c.$$

Applying $y(1) = 1$ implies $c = 1$, giving the solution

$$y^2 + x^2 = 2.$$

10.1.3 INTEGRATING FACTOR

To solve

$$\frac{dy}{dx} + P(x)y = f(x)$$

calculate the integrating factor

$$R(x) = \exp\left(\int P(x)\,dx\right)$$

then

$$\begin{aligned}\frac{d}{dx}(R(x)y) &= R(x)f(x) \\ \implies y &= \frac{1}{R(x)}\left(\int R(x)f(x)\,dx + c\right)\end{aligned}$$

where c is the integrating constant.

EXAMPLES

1. To solve $x\dfrac{dy}{dx} + 4y = x^3 - x$, for $x > 0$, the equation is divided through by x to give

$$\frac{dy}{dx} + \frac{4}{x}y = x^2 - 1.$$

The integrating factor is then

$$\begin{aligned}R(x) &= \exp\left(\int \frac{4}{x}\,dx\right) \\ &= e^{4\ln x} \\ &= x^4.\end{aligned}$$

The differential equation then becomes

$$\begin{aligned}\frac{d}{dx}(x^4 y) &= x^4(x^2 - 1) \\ \implies x^4 y &= \int (x^6 - x^4)\,dx \\ &= \frac{x^7}{7} - \frac{x^5}{5} + c\end{aligned}$$

so

$$y = \frac{x^3}{7} - \frac{x}{5} + \frac{c}{x^4}.$$

10.2 SECOND ORDER DIFFERENTIAL EQUATIONS

10.2.1 HOMOGENEOUS

A second order homogeneous equation with constant coefficients,

$$ay'' + by' + cy = 0$$

is solved with the substitution $y = e^{mx}$. The differential equation becomes

$$\begin{aligned} am^2e^{mx} + bme^{mx} + ce^{mx} &= 0 \\ \implies \quad e^{mx}(am^2 + bm + c) &= 0 \\ \implies \quad am^2 + bm + c &= 0. \end{aligned}$$

This is the **characteristic equation** and has solutions $m = m_1, m_2$. The form of the general solution depends on m_1 and m_2 — the roots of the characteristic equation. There are **three** cases:

1. The two roots are real and $m_1 \neq m_2$, then

$$y(x) = c_1e^{m_1x} + c_2e^{m_2x}.$$

2. The two roots are real and $m_1 = m_2$, then

$$y(x) = c_1e^{m_1x} + c_2xe^{m_1x}.$$

3. The roots are complex, $m_1 = \alpha + i\beta$ and $m_2 = \alpha - i\beta$, then

$$y(x) = e^{\alpha x}\left(c_1\cos(\beta x) + c_2\sin(\beta x)\right),$$

where c_1, c_2 are arbitrary constants in all cases.

EXAMPLES

1. Solve $y'' - y' - 12y = 0$ for $y(x)$.

The characteristic equation for the differential equation is

$$m^2 - m - 12 = 0$$

which factorises easily to

$$(m - 4)(m + 3) = 0$$

giving two real roots,

$$m_1 = 4 \qquad \text{and} \qquad m_2 = -3.$$

The general solution is therefore

$$y(x) = c_1 e^{4x} + c_2 e^{-3x}.$$

2. Solve $4\dfrac{d^2y}{dx^2} - 12\dfrac{dy}{dx} + 9y = 0.$

The characteristic equation is

$$4m^2 - 12m + 9 = 0$$

which factorises to

$$(2m - 3)^2 = 0$$

implying a repeated root $m_1 = \dfrac{3}{2}$. The general solution is then of the form

$$y(x) = c_1 e^{3x/2} + c_2 x e^{3x/2}.$$

> There are two common differential equations that, together with their solutions, are worth considering:
>
> $$\begin{aligned} y'' + \lambda^2 y = 0 &\quad\Longrightarrow\quad y = A\sin(\lambda x) + B\cos(\lambda x) \\ y'' - \lambda^2 y = 0 &\quad\Longrightarrow\quad y = A\sinh(\lambda x) + B\cosh(\lambda x) \\ &\quad\quad\text{or}\quad\quad y = c_1 e^{\lambda x} + c_2 e^{-\lambda x}. \end{aligned}$$

EXAMPLES

1. Solve $y'' + 8y = 0$.

The characteristic equation is

$$m^2 + 8 = 0$$

immediately giving the complex roots

$$m_1 = 2\sqrt{2}i \qquad \text{and} \qquad m_2 = -2\sqrt{2}i$$

and the general solution is

$$y(x) = c_1 \cos(2\sqrt{2}x) + c_2 \sin(2\sqrt{2}x).$$

2. The equation $y''(x) = 4y(x)$, with $y(0) = 1, y(\infty) = 0$ is solved by writing the general solution

$$y(x) = c_1 e^{2x} + c_2 e^{-2x}$$

where c_1, c_2 are arbitrary constants easily found by application of the boundary conditions so that

$$y(x) = e^{-2x}.$$

Note that the exponential form of the general solution is used since the boundary conditions extend to infinity.

3. The equation $y''(x) = 4y(x)$, with $y(0) = 0, y(1) = 1$ is solved by writing the general solution

$$y(x) = c_1 \sinh 2x + c_2 \cosh 2x$$

where c_1, c_2 are arbitrary constants easily found by application of the boundary conditions so that

$$y(x) = \frac{\sinh 2x}{\sinh 2}.$$

Note that the hyperbolic form of the general solution is used since the boundary conditions are finite.

10.2.2 INHOMOGENEOUS

The differential equation

$$ay'' + by' + cy = F(x)$$

has the solution

$$y(x) = y_h(x) + y_p(x)$$

where $y_h(x)$ is the *general* solution to the homogeneous equation and $y_p(x)$ is a *particular* solution to the complete equation.

EXAMPLE

The differential equation $\dfrac{d^2y}{dx^2} - y = x$ has solution

$$y(x) = c_1 e^{x} + c_2 e^{-x} - x$$

where $-x$ is the particular solution.

The **method of undetermined coefficients** is used where $F(x)$ is of the form

$$e^{rx}, \quad \cos kx, \quad \sin kx, \quad ax^2 + bx + c.$$

The *form* of $y_p(x)$ is "guessed" and then the constants determined by substituting the guess into the differential equation and equating the coefficients:

RHS Forcing Term	Try
e^{rx}	Ae^{rx}
$\cos kx$	$A\cos kx + B\sin kx$
$\sin kx$	$A\cos kx + B\sin kx$
$a_nx^n + \cdots + a_0$	$A_nx^n + \cdots + A_0$

EXAMPLES

1. Solve $\dfrac{d^2y}{dx^2} + 6\dfrac{dy}{dx} + 8y = 10x$.

The homogeneous solution is

$$y_h(x) = c_1e^{-2x} + c_2e^{-4x}$$

and we guess the form of $y_p(x)$ to be $Ax + B$ since $F(x)$ is a polynomial of degree 1. Substituting gives,

$$6A + 8Ax + 8B = 10x$$

so that equating coefficients yields two equations to be solved for two unknowns:

$$\begin{aligned} 8A &= 10 \\ 6A + 8B &= 0, \end{aligned}$$

giving the solution $A = \dfrac{5}{4}$ and $B = -\dfrac{15}{16}$ so that the complete solution is

$$y(x) = c_1e^{-2x} + c_2e^{-4x} + \frac{5}{4}x - \frac{15}{16}.$$

2. Solve $y'' + y = \sin 2x$.

The solution to the homogeneous equation is

$$y_h(x) = c_1\cos x + c_2\sin x.$$

Trying the form of a particular solution as

$$y_p(x) = A\sin(2x) + B\cos(2x)$$

leads to the set of equations

$$-4A + A = 1$$
$$-4B + B = 0\ ,$$

which gives $A = -\frac{1}{3}$ and $B = 0$. The complete solution is thus

$$y(x) = c_1 \cos x + c_2 \sin x - \frac{1}{3} \sin 2x.$$

3. Solve $y'' + y = \sin x$.

   As with the previous problem

$$y_h(x) = c_1 \cos x + c_2 \sin x$$

however a guess of $A \sin x + B \cos x$ will not work because $\sin x$ and $\cos x$ are already solutions to the homogeneous equation. In this case try

$$y_p(x) = Ax \sin x + Bx \cos x + C \sin x + D \cos x$$

which eventually gives the solution

$$y(x) = c_1 \cos x + c_2 \sin x + \frac{1}{2} \sin x - \frac{1}{2} x \cos x.$$

The extra term $x \cos x$ grows in amplitude and is the **resonance** term.

4. Solve $y'' - 2y' + y = 4e^{3x}$.

   The solution to the homogeneous equation is

$$y_h(x) = c_1 e^x + c_2 x e^x$$

with the extra x factor in xe^x coming from the repeated root in the characteristic equation. The form of a particular solution is

$$y_p(x) = Ae^{3x}$$

leading to $A = 1$. Hence the solution is

$$y(x) = c_1 e^x + c_2 x e^x + e^{3x}.$$

10.3 EXAMPLE QUESTIONS

(Answers are given in Chapter 14)

1. Solve for $y = y(x)$ by direct integration:

   (i) $\dfrac{dy}{dx} = (x+1)^2$

   (ii) $\dfrac{dy}{dx} = \dfrac{1}{x^2}$

   (iii) $(x+1)\dfrac{dy}{dx} = x$

   (iv) $\dfrac{dx}{dy} = 5x$

   (v) $\dfrac{dy}{dx} = -3x\sin x$

2. Solve for $y(x)$ using separation methods:

   (i) $y^3\dfrac{dy}{dx} = 1$

   (ii) $\dfrac{dy}{dx} = \dfrac{y^3}{x^2}$

   (iii) $\dfrac{dy}{dx} = \dfrac{y+1}{x}$

   (iv) $\dfrac{dy}{dx} = e^{3x+2y}$

   (v) $(y - yx^2)\dfrac{dy}{dx} = (y+1)^2$

   (vi) $\dfrac{dx}{dy} = \dfrac{1+2y^2}{y\sin x}$

   (vii) $\dfrac{dy}{dx} = \dfrac{x\sin x\, e^{-y}}{y}$

   (viii) $\dfrac{dx}{dy} = \left(\dfrac{2y+3}{4x+5}\right)^2$

3. Solve for $y(x)$ using an integrating factor:

   (i) $2\dfrac{dy}{dx} + 10y = 1$

   (ii) $x\dfrac{dy}{dx} + 2y = 3$

   (iii) $\dfrac{dy}{dx} + 2xy = x$

   (iv) $y' + 3x^2y = x^2$

   (v) $\cos x\dfrac{dy}{dx} + y\sin x = 1$

   (vi) $(1 - x^3)\dfrac{dy}{dx} = 3x^2y$

4. Solve using any workable method:

   (i) $\dfrac{dQ}{dt} = k(Q - 70)$, for $k =$ a constant.

   (ii) $(1 + e^x)\dfrac{dy}{dx} + e^x y = 0$

   (iii) $\dfrac{dr}{d\theta} + r\sec\theta = \cos\theta$

   (iv) $\dfrac{dP}{dt} = P(1-P)$

   (v) $2\dfrac{dy}{dx} - \dfrac{1}{y} = \dfrac{2x}{y}$

   (vi) $L\dfrac{di}{dt} + Ri = E$, with L, R, E constants.

5. Solve for $y = y(x)$:

   (i) $y'' - 16y = 0$

   (ii) $y'' + 4y' - y = 0$

   (iii) $12y'' - 5y' - 2y = 0$

   (iv) $y'' + 9y = 0$

   (v) $\dfrac{d^2y}{dx^2} + 8\dfrac{dy}{dx} + 16y = 0$

   (vi) $8y'' + 2y' - y = 0$

   (vii) $2y'' + 5y' = 0$

   (viii) $3y'' + 2y' + y = 0$

   (ix) $\dfrac{d^2y}{dx^2} - 10\dfrac{dy}{dx} + 25y = 0$

6. Solve for $y = y(x)$:

   (i) $\dfrac{d^2y}{dx^2} - 3\dfrac{dy}{dx} - 10y = -3$

   (ii) $\dfrac{d^2y}{dx^2} + 2\dfrac{dy}{dx} + y = x^2$

   (iii) $\dfrac{d^2y}{dx^2} + y = \cos(3x)$

   (iv) $\dfrac{d^2y}{dx^2} + y = e^{2x}$

   (v) $\dfrac{d^2y}{dx^2} + 2\dfrac{dy}{dx} + y = 6\sin(2x)$

   (vi) $\dfrac{d^2y}{dx^2} - y = e^x + x^2$

CHAPTER 11

MULTIVARIABLE CALCULUS

11.1 PARTIAL DIFFERENTIATION

Given $f = f(x, y)$ then $\frac{\partial f}{\partial x}$ is calculated by treating y as a constant while differentiating f with respect to x and $\frac{\partial f}{\partial y}$ is obtained by treating x as a constant and differentiating with respect to y. All higher derivatives are treated similarly.

EXAMPLES

1. If $f(x, y) = x^2 + xy + y^2$ then

$$\frac{\partial f}{\partial x} = 2x + y, \qquad \frac{\partial f}{\partial y} = x + 2y, \qquad \frac{\partial^2 f}{\partial x \partial y} = 1.$$

2. If $f(x, y) = \sin x + x \cos 2y$ then

$$\frac{\partial f}{\partial x} = \cos x + \cos 2y, \quad \frac{\partial f}{\partial y} = -2x \sin 2y, \quad \frac{\partial^2 f}{\partial x \partial y} = -2 \sin 2y.$$

3. If $f = x^3 y^5$ then

$$\frac{\partial f}{\partial x} = 3x^2 y^5, \quad \frac{\partial^2 f}{\partial x^2} = 6xy^5, \quad \frac{\partial f}{\partial y} = 5x^3 y^4,$$

$$\frac{\partial^2 f}{\partial y^2} = 20x^3 y^3, \quad \frac{\partial^2 f}{\partial x \partial y} = 15x^2 y^4.$$

If $z = f(x, y)$ is the height as a function of x, y then $\frac{\partial f}{\partial x}$ is the slope in the x direction and $\frac{\partial f}{\partial y}$ is the slope in the y direction.

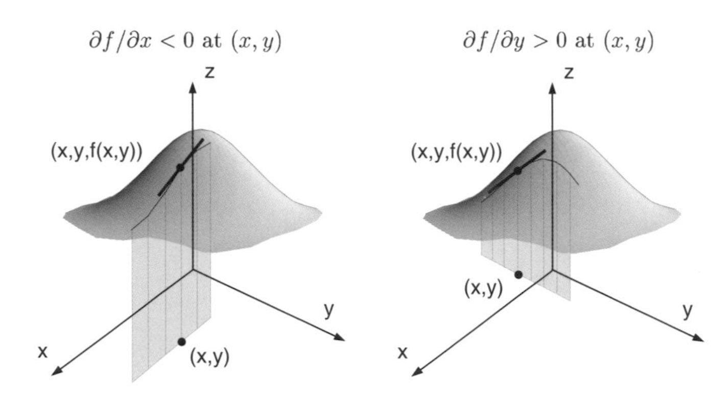

11.2 GRAD, DIV AND CURL

The *gradient* vector, call grad, del, or nabla, denoted by $\underset{\sim}{\boldsymbol{\nabla}}$, is a vector operator:

$$\underset{\sim}{\boldsymbol{\nabla}} = \left(\frac{\partial}{\partial x}, \frac{\partial}{\partial y}, \frac{\partial}{\partial z}\right).$$

If $\underset{\sim}{\boldsymbol{v}} = (v_1, v_2, v_3)$ and $f = f(x, y, z)$ then

$$\begin{aligned}
\underset{\sim}{\boldsymbol{\nabla}} f &= \left(\frac{\partial f}{\partial x}, \frac{\partial f}{\partial y}, \frac{\partial f}{\partial z}\right) \\
\underset{\sim}{\boldsymbol{\nabla}} \cdot \underset{\sim}{\boldsymbol{v}} &= \frac{\partial v_1}{\partial x} + \frac{\partial v_2}{\partial y} + \frac{\partial v_3}{\partial z} \qquad \text{the divergence, or div, of } \underset{\sim}{\boldsymbol{v}} \\
\underset{\sim}{\boldsymbol{\nabla}} \times \underset{\sim}{\boldsymbol{v}} &= \begin{vmatrix} \underset{\sim}{\boldsymbol{i}} & \underset{\sim}{\boldsymbol{j}} & \underset{\sim}{\boldsymbol{k}} \\ \frac{\partial}{\partial x} & \frac{\partial}{\partial y} & \frac{\partial}{\partial z} \\ v_1 & v_2 & v_3 \end{vmatrix} \qquad \text{the curl of } \underset{\sim}{\boldsymbol{v}} \\
\underset{\sim}{\boldsymbol{\nabla}}^2 f &= \underset{\sim}{\boldsymbol{\nabla}} \cdot \underset{\sim}{\boldsymbol{\nabla}} f = \frac{\partial^2 f}{\partial x^2} + \frac{\partial^2 f}{\partial y^2} + \frac{\partial^2 f}{\partial z^2} \qquad \text{the Laplacian.}
\end{aligned}$$

EXAMPLES

1. If $f = 2x^2y + z^3$ then

$$\underset{\sim}{\nabla} f = \frac{\partial f}{\partial x}\underset{\sim}{i} + \frac{\partial f}{\partial y}\underset{\sim}{j} + \frac{\partial f}{\partial z}\underset{\sim}{k}$$
$$= 4xy\underset{\sim}{i} + 2x^2\underset{\sim}{j} + 3z^2\underset{\sim}{k}$$
$$\equiv (4xy, 2x^2, 3z^2).$$

2. If $f = x + xyz$ then

$$\underset{\sim}{\nabla} f = \left(\frac{\partial f}{\partial x}, \frac{\partial f}{\partial y}, \frac{\partial f}{\partial z}\right)$$
$$= (1 + yz, xz, xy).$$

3. If $\underset{\sim}{v} = xyz\,\underset{\sim}{i} + 3y\,\underset{\sim}{j} + \sin x\,\underset{\sim}{k}$ then

$$\underset{\sim}{\nabla} \cdot \underset{\sim}{v} = \frac{\partial}{\partial x}[xyz] + \frac{\partial}{\partial y}[3y] + \frac{\partial}{\partial z}[\sin x]$$
$$= yz + 3$$

and

$$\underset{\sim}{\nabla} \times \underset{\sim}{v} = \begin{vmatrix} \underset{\sim}{i} & \underset{\sim}{j} & \underset{\sim}{k} \\ \frac{\partial}{\partial x} & \frac{\partial}{\partial y} & \frac{\partial}{\partial z} \\ xyz & 3y & \sin x \end{vmatrix}$$
$$= \left(\frac{\partial}{\partial y}[\sin x] - \frac{\partial}{\partial z}[3y]\right)\underset{\sim}{i} - \left(\frac{\partial}{\partial x}[\sin x] - \frac{\partial}{\partial z}[xyz]\right)\underset{\sim}{j}$$
$$+ \left(\frac{\partial}{\partial x}[3y] - \frac{\partial}{\partial y}[xyz]\right)\underset{\sim}{k}$$
$$= 0\underset{\sim}{i} - (\cos x - xy)\underset{\sim}{j} - xz\underset{\sim}{k}$$
$$= (0, xy - \cos x, -xz).$$

4. If $f = x^2 + y^2 + yz^2$ then

$$\underset{\sim}{\nabla}^2 f = \frac{\partial^2 f}{\partial x^2} + \frac{\partial^2 f}{\partial y^2} + \frac{\partial^2 f}{\partial z^2}$$
$$= 2 + 2 + 2y.$$

5. If $f = x + y + z^2$ then $\underset{\sim}{\nabla} f = (1, 1, 2z)$ and $\underset{\sim}{\nabla}^2 f = 2$.

6. If $\underset{\sim}{v} = (x, x, z + x)$ then $\underset{\sim}{\nabla} \cdot \underset{\sim}{v} = 2$ and $\underset{\sim}{\nabla} \times \underset{\sim}{v} = (0, -1, 1)$.

The slope of $z = f(x, y)$ in the direction given by the **unit vector** $\underset{\sim}{\boldsymbol{u}}$ is

$$f_{\underset{\sim}{\boldsymbol{u}}} = \underset{\sim}{\boldsymbol{u}} \cdot \underset{\sim}{\boldsymbol{\nabla}} f.$$

EXAMPLES

1. If $z = f(x, y) = 10 - x^2 - y^2$ then $\underset{\sim}{\boldsymbol{\nabla}} f = (-2x, -2y)$. At the point $(1, 2)$, $\underset{\sim}{\boldsymbol{\nabla}} f = (-2, -4)$. To find the slope in the direction of $(1, 1)$ find

$$\frac{(1,1)}{\sqrt{2}} \cdot (-2, -4) = \frac{-6}{\sqrt{2}}.$$

2. If $z = 1 - x + y^2$ then at the point $(3, 1)$ the slope in the direction $(3, 4)$ is

$$\begin{aligned} \frac{(3,4)}{5} \cdot (-1, 2y)|_{(3,1)} &= \frac{(3,4)}{5} \cdot (-1, 2) \\ &= \frac{-3+8}{5} = 1. \end{aligned}$$

If $z = f(x, y)$ then the gradient vector $\underset{\sim}{\boldsymbol{\nabla}} f = \left(\frac{\partial f}{\partial x}, \frac{\partial f}{\partial y}\right)$ has the properties

(i) $\underset{\sim}{\boldsymbol{\nabla}} f$ points in the direction of maximum slope.

(ii) $\underset{\sim}{\boldsymbol{\nabla}} f$ is perpendicular to the contours (lines of constant z).

(iii) $||\underset{\sim}{\boldsymbol{\nabla}} f||$ is the magnitude of the maximum gradient.

EXAMPLES

1. If $z = y^2 - x$ is the height of a hill then $\underset{\sim}{\boldsymbol{\nabla}} f = (-1, 2y)$. If we are at the point $(1, 3)$ then

   (i) $\underset{\sim}{\boldsymbol{\nabla}} f = (-1, 6)$ is the direction to move uphill fastest.

   (ii) The slope in that direction is $||(-1, 6)|| = \sqrt{1^2 + 6^2} = \sqrt{37}$.

   (iii) The contours are in the direction $\pm(6, 1)$ since $(6, 1) \cdot (-1, 6) = 0$.

2. If $T = 1 - x + y^2$ is the temperature on a flat plate then $\underset{\sim}{\nabla} T = (-1, 2y)$. At the point $(3, 1)$ the direction that increases temperature fastest is $(-1, 2)$. The direction in which temperature stays the same is $\pm(2, 1)$.

If $z = f(x, y)$ but $x(t)$ and $y(t)$ are functions of t the **chain rule** for differentiation of f with respect to t is

$$\frac{df}{dt} = \frac{\partial f}{\partial x}\frac{dx}{dt} + \frac{\partial f}{\partial y}\frac{dy}{dt} = \underset{\sim}{\nabla} f \cdot \underset{\sim}{v}, \quad \text{where } \underset{\sim}{v} = \left(\frac{dx}{dt}, \frac{dy}{dt}\right).$$

EXAMPLE

If $p = x^2 + y^2$ and $x = t^2$, $y = t^3$ then to find dp/dt without substitution we get:

$$\frac{dp}{dt} = \frac{dp}{dx}\frac{dx}{dt} + \frac{dp}{dy}\frac{dy}{dt} = 2x2t + 2y3t^2 = 4t^3 + 6t^5.$$

11.3 DOUBLE INTEGRALS

If R is defined by $a \le x \le b$, $g_1(x) \le y \le g_2(x)$ with g_1 and g_2 continuous on $[a, b]$ then

$$\iint_R f(x, y)\, dA = \int_a^b \int_{g_1(x)}^{g_2(x)} f(x, y)\, dy dx.$$

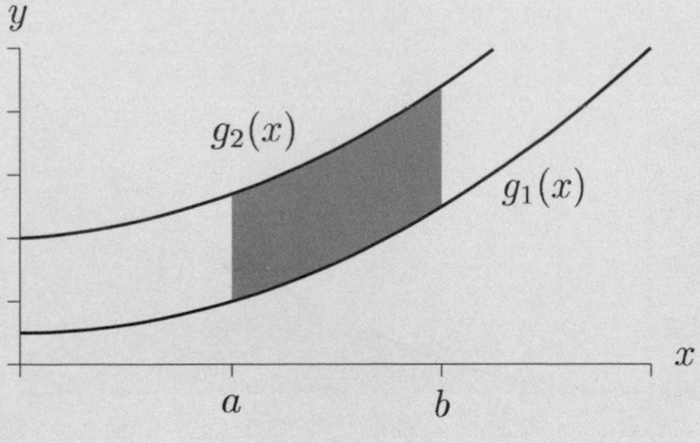

The double integral represents the volume under the surface $z = f(x, y)$ over the region R.

EXAMPLES

1. If R is the region given by $0 \le x \le 3,\ \ 0 \le y \le 1$ and $f(x,y) = xy$ then

$$\begin{aligned}
\iint_R xy\,dA &= \int_0^1 \int_0^3 xy\,dxdy \\
&= \int_0^1 \left[\frac{1}{2}x^2 y\right]_0^3 dy \\
&= \int_0^1 \frac{9}{2}y\,dy \\
&= \left[\frac{9}{4}y^2\right]_0^1 \\
&= \frac{9}{4}.
\end{aligned}$$

2. If R is the region given by $0 < x < 1$ and $0 < y < x^2$ then for $f(x,y) = x + 3y^2$ we have

$$\begin{aligned}
\iint_R x + 3y^2 &= \int_0^1 \int_0^{x^2} x + 3y^2\,dy\,dx \\
&= \int_0^1 \left[xy + y^3\right]_0^{x^2} dx \\
&= \int_0^1 x^3 + x^6\,dx \\
&= \left[\frac{1}{4}x^4 + \frac{1}{7}x^7\right]_0^1 \\
&= \frac{1}{4} + \frac{1}{7} = \frac{11}{28}.
\end{aligned}$$

The region R is drawn below.

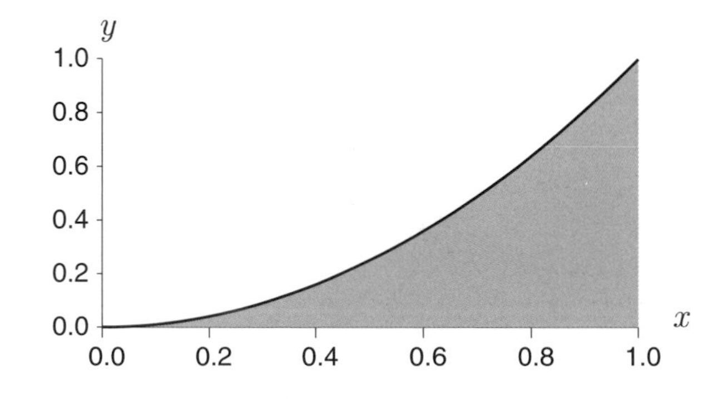

3. Consider the region R given by

$$0 \le x \le 1, \qquad 0 \le y \le x :$$

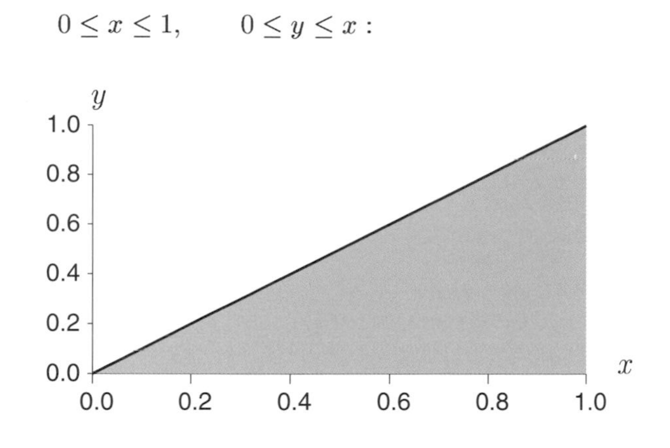

If $f(x, y) = 2y + x^3$ then

$$\begin{aligned}
\iint_R 2y + x^3 \, dA &= \int_{x=0}^{1} \int_{y=0}^{x} 2y + x^3 \, dy dx \\
&= \int_0^1 \left[y^2 + yx^3\right]_0^x \, dx \\
&= \int_0^1 x^2 + x^4 \, dx \\
&= \left[\frac{1}{3}x^3 + \frac{1}{5}x^5\right]_0^1 \\
&= \frac{1}{3} + \frac{1}{5} = \frac{8}{15}.
\end{aligned}$$

We can also reverse the order of integration, doing the x integral first:

$$\begin{aligned}
\iint_R 2y + x^3 \, dA &= \int_{y=0}^{1} \int_{x=y}^{1} 2y + x^3 \, dx dy \\
&= \int_{y=0}^{1} \left[2xy + \frac{1}{4}x^4\right]_y^1 \, dy \\
&= \int_{y=0}^{1} 2y + \frac{1}{4} - 2y^2 - \frac{1}{4}y^4 \, dy \\
&= \left[y^2 + \frac{1}{4}y - \frac{2}{3}y^3 - \frac{1}{20}y^5\right]_0^1 = \frac{8}{15}.
\end{aligned}$$

11.4 EXAMPLE QUESTIONS

(Answers are given in Chapter 14)

1. Find $\frac{\partial f}{\partial x}$, $\frac{\partial f}{\partial y}$ and $\frac{\partial^2 f}{\partial x \partial y}$ if

   (i) $f(x,y) = y^2 + 2xy + 1$

   (ii) $f(x,y) = \sin(xy)$

   (iii) $f(x,y) = x^2 + \cos(x+y)$

   (iv) $f(x,y) = e^{x-y}$

   (v) $f(x,y) = yx^3$

2. Evaluate

   (i) $\underset{\sim}{\nabla} f$ where $f = x^2 - y^2 + 2xz$

   (ii) $\underset{\sim}{\nabla} \phi$ where $\phi = \tan(xyz)$

   (iii) $\underset{\sim}{\nabla} \cdot \underset{\sim}{v}$ where $\underset{\sim}{v} = (x^2, \sqrt{y}, 9xz)$

   (iv) $\underset{\sim}{\nabla} \times \underset{\sim}{u}$ where $\underset{\sim}{u} = (xyz, xyz, xyz)$

   (v) $\underset{\sim}{\nabla} \cdot (x, 2y, 3z)$

   (vi) $\underset{\sim}{\nabla} \cdot (x^2, 2y^2, 3z^2)$

   (vii) $\underset{\sim}{\nabla}^2 (x^2 + y^2 + z^2)$

   (viii) $\underset{\sim}{\nabla}^2 (xyz)^2$

   (ix) $\underset{\sim}{\nabla} \times (y^2, x^2, z^2)$

   (x) $\underset{\sim}{\nabla} \times (z^2, x^2, y^2)$

3. For the functions $z = f(x,y) = xy^2 - y$ find the following at the point $(1,2)$.

   (i) The direction of maximum gradient.

   (ii) The magnitude of the maximum slope.

   (iii) The direction of the contours.

   (iv) The slope in the direction $(1,1)$.

4. Repeat the previous question for $z = 1 - x - e^y$ at the point $(1,0)$.

5. If $f = x^2 - y$ but $x = t^2$ and $y = t^3$ find $\frac{df}{dt}$ using the chain rule.

6. Evaluate:

   (i) $\displaystyle\int_0^9 \int_{x^2/9}^{3\sqrt{x}} 2 \, dy dx$

   (ii) $\displaystyle\int_0^{16} \int_{x/4}^{\sqrt{x}} y \, dy dx$

   (iii) $\displaystyle\int_0^6 \int_0^4 (x^2 + y^2) \, dy dx$

7. Find $\iint_R x^2 + y^2 \, dA$ where R is the region between the line $y = x^2$ and $y = 1$ shown in the diagram below.

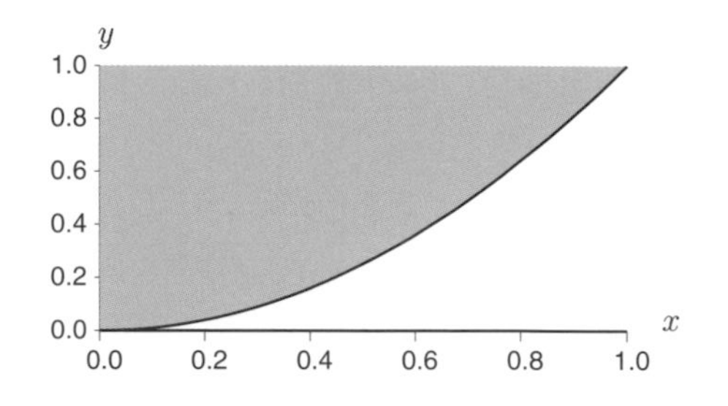

8. Repeat the above question by reversing the order of integration.

9. Find $\iint x + y \, dA$ where R is the region is between the line $y = \sqrt{x}$ and $y = x$ shown in the diagram below.

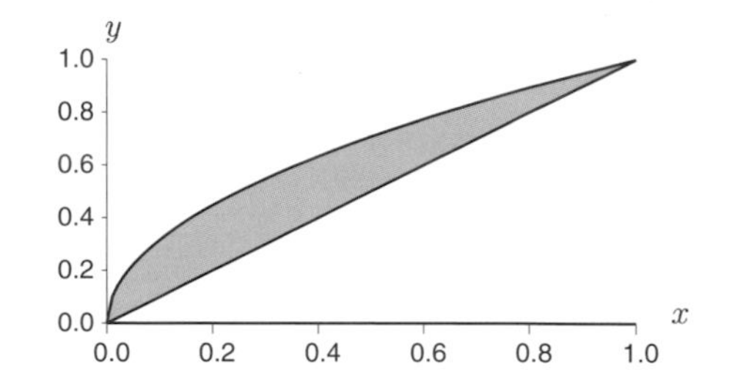

10. Repeat the above question by reversing the order of integration.

11. Show that for any vector $\underset{\sim}{v}$

$$\underset{\sim}{\nabla} \cdot (\underset{\sim}{\nabla} \times \underset{\sim}{v}) = \underset{\sim}{0}.$$

12. Show that

$$c(x,t) = \frac{1}{\sqrt{t}} e^{-x^2/(4t)}$$

satisfies the diffusion equation

$$\frac{\partial^2 c}{\partial x^2} = \frac{\partial c}{\partial t}.$$

CHAPTER 12

NUMERICAL SKILLS

12.1 INTEGRATION

An integral $\int_a^b f(x)\,dx$ can be calculated by estimating the area under the curve as a series of rectangles.

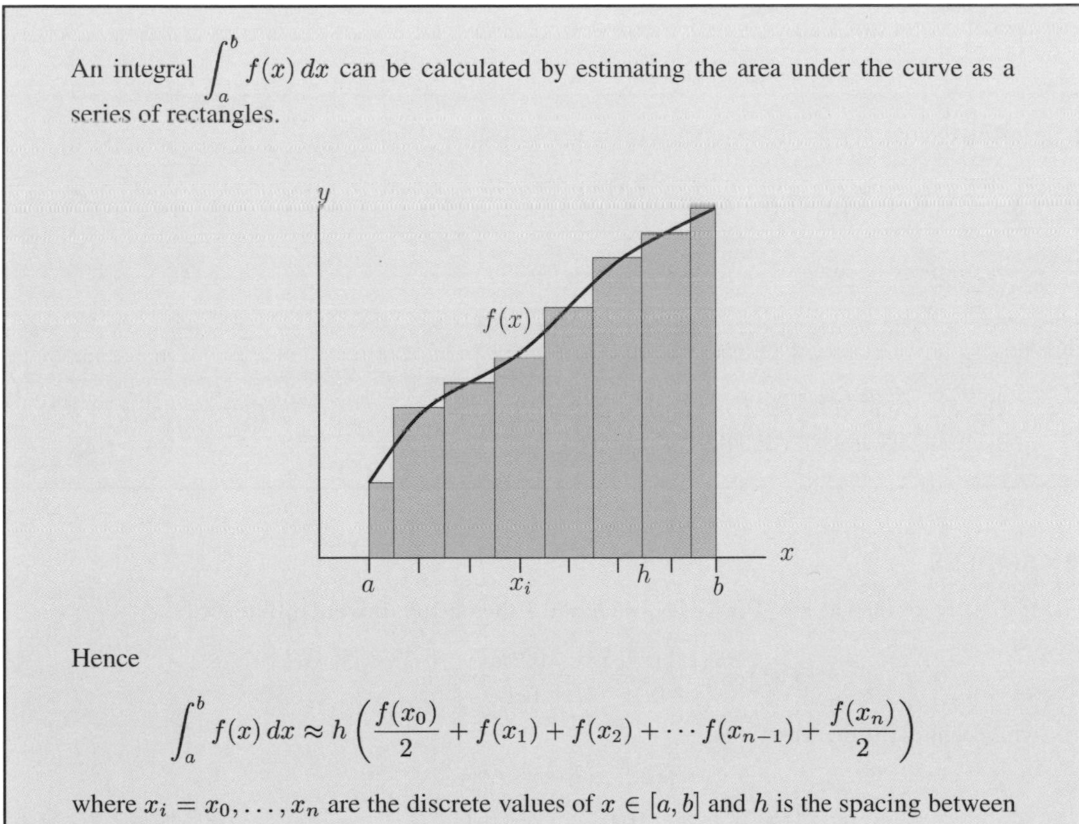

Hence

$$\int_a^b f(x)\,dx \approx h\left(\frac{f(x_0)}{2} + f(x_1) + f(x_2) + \cdots f(x_{n-1}) + \frac{f(x_n)}{2}\right)$$

where $x_i = x_0, \ldots, x_n$ are the discrete values of $x \in [a, b]$ and h is the spacing between points: $h = x_i - x_{i-1}$.

EXAMPLES

1. If $f(x) = x^2$ then for $x \in [0,1]$ and five equally spaced points $x_0 = 0$, $x_1 = 0.25$, $x_2 = 0.5$, $x_3 = 0.75$, $x_4 = 1.0$ so $h = 0.25$ and

$$\begin{aligned}\int_0^1 x^2\,dx &\approx 0.25\left(\frac{0}{2} + (0.25)^2 + (0.5)^2 + (0.75)^2 + \frac{1^2}{2}\right)\\ &\approx 0.34.\end{aligned}$$

   Note that the exact answer is $1/3$.

2. If $f(x) = x$ then

$$\begin{aligned}\int_1^2 x\,dx &\approx 0.1\left(\frac{1}{2} + 1.1 + 1.2 + \cdots + 1.9 + \frac{2}{2}\right)\\ &= 1.5.\end{aligned}$$

   Note that the exact answer is also 1.5.

12.2 DIFFERENTIATION

A derivative can be approximated using the derivative definition

$$f'(x) \approx \frac{f(x+h) - f(x)}{h}$$

or for more accuracy

$$f'(x) \approx \frac{f(x+h) - f(x-h)}{2h}.$$

These are called forward and central difference respectively. The smaller the value of h the more accuracy the result.

EXAMPLES

1. If $f(x) = x^2$ then at $x = 1$ if we choose $h = 0.1$ then using forward differences

$$f'(1) \approx \frac{(1.1)^2 - 1^2}{0.1} = \frac{0.21}{0.1} = 2.1$$

   while central differences gives

$$f'(1) \approx \frac{(1.1)^2 - (0.9)^2}{0.2} = \frac{0.21}{0.1} = 2.$$

   The exact answer is $f'(1) = 2$.

2. If $f(x) = e^x$ then using central differences at $x = 0$ with $h = 0.2$ gives

$$f'(0) \approx \frac{e^{0.2} - e^{-0.2}}{0.02} \approx 1.0067$$

which compares well with the exact answer $f'(0) = 1$.

12.3 NEWTON'S METHOD

To find a zero of a function $f(x)$ guess a starting answer and then iterate using the formula:

$$x_{n+1} = x_n - \frac{f(x_n)}{f'(x_n)}$$

where $x_0, x_1, \ldots, x_n, \ldots$ are successive approximations to the zero.

EXAMPLES

1. To find x such that $f(x) = x^2 - 2 = 0$ first find $f'(x) = 2x$ then guess an answer $x_0 = 1$. The next approximation is

$$\begin{aligned} x_1 &= x_0 - \frac{f(x_0)}{f'(x_0)} \\ &= 1 - \frac{1^2 - 2}{2 \times 1} \\ &= 1 + \frac{1}{2} = 1.5. \end{aligned}$$

The second approximation is

$$x_2 = 1.5 - \frac{(1.5)^2 - 2}{3} = 1.416.$$

This can be continued to whatever accuracy is required. The exact answer is $\sqrt{2} \approx 1.4142$.

2. To find a zero of $f(x) = x^3 - x - 1$ find $f'(x) = 3x^2 - 1$ then use $x_0 = 1$ as a starting guess. Hence

$$\begin{aligned} x_1 &= 1 - \frac{-1}{2} = 1.5 \\ x_2 &= 1.5 - \frac{1.5^3 - 1.5 - 1}{3(1.5)^3 - 1} \approx 1.348 \\ x_3 &= 1.348 - \frac{1.348^3 - 1.348 - 1}{3(1.348)^3 - 1} \approx 1.325. \end{aligned}$$

The exact answer is approximately 1.324718.

12.4 DIFFERENTIAL EQUATIONS

To solve the differential equation

$$\frac{dy}{dx} = f(x, y), \quad f(x_0) = y_0$$

use the forward derivative differencing formula so

$$y_{n+1} = y_n + hf(x_n, y_n)$$

where x_n are the discrete values of x, y_n the solutions $y(x_n)$ and $h = x_{n+1} - x_n$.

EXAMPLES

1. To solve $dy/dx = x + y$ with $y(0) = 1$ consider a series of x values $0, 0.1, 0.2, 0.3 \ldots$ with $h = 0.1$. Thus $x_0 = 0$, $y_0 = 1$

$$\begin{aligned} y_1 &= 1 + 0.1f(0, 1) = 1 + 0.1(0 + 1) = 1.1 \\ y_2 &= 1.1 + 0.1f(0.1, 1.1) = 1.22 \\ y_3 &= 1.22 + 0.1f(0.2, 1.22) = 1.362 \\ y_4 &= 1.362 + 0.1f(0.3, 1.362) = 1.5282. \end{aligned}$$

Thus we have an approximate solution set of point $(0, 1)$, $(0.1, 1.1)$, $(0.2, 1.22)$, $(0.3, 1.362)$, $(0.4, 1.5282)$ representing the solution $y = y(x)$. The exact answer is $y(x) = -x - 1 + 2e^x$ so that $y(0.1) \approx 1.11$, $y(0.2) \approx 1.24$ and so on.

2. To find $y(0.4)$ using $\frac{dy}{dx} = x^2 + 1$ with $y(0) = 0$ use $h = 0.4$ so that

$$y(0.4) \approx y(0) + 0.4(0^2 + 1) = 0.4.$$

To find a better approximation use $h = 0.2$ so that

$$\begin{aligned} y(0.2) &= 0 + 0.2(0^2 + 1) = 0.2 \\ y(0.4) &= 0.2 + 0.2(0.2^2 + 1) = 0.408. \end{aligned}$$

The exact answer is $y(x) = \frac{x^3}{3} + x$ giving $y(0.4) \approx 0.42$.

12.5 FOURIER SERIES

A function $f(x)$, $x \in [-p, p]$ can be expressed as a series of sine and cosine terms:

$$f(x) = \frac{a_0}{2} + \sum_{n=1}^{\infty} \left(a_n \cos \frac{n\pi}{p} x + b_n \sin \frac{n\pi}{p} x \right)$$

where the coefficients a_n, b_n are given by

$$\begin{aligned} a_0 &= \frac{1}{p} \int_{-p}^{p} f(x)\, dx \\ a_n &= \frac{1}{p} \int_{-p}^{p} f(x) \cos \frac{n\pi}{p} x\, dx \\ b_n &= \frac{1}{p} \int_{-p}^{p} f(x) \sin \frac{n\pi}{p} x\, dx. \end{aligned}$$

EXAMPLE

The function $f(x) = \begin{cases} 0, & -1 \le x < 0 \\ 1, & 0 \le x \le 1 \end{cases}$ has Fourier coefficients

$$\begin{aligned} a_0 &= \int_{-1}^{1} f(x)\, dx = \int_{0}^{1} 1\, dx = 1 \\ a_n &= \int_{0}^{1} 1 \cos n\pi x\, dx = \frac{\sin n\pi}{n\pi} = 0 \\ b_n &= \int_{0}^{1} 1 \sin n\pi x\, dx = \frac{1 - \cos n\pi}{n\pi} = \frac{1 - (-1)^n}{n\pi} \end{aligned}$$

so $f(x) = \frac{1}{2} + \sum_{n=1}^{\infty} \frac{1 - (-1)^n}{n\pi} \sin n\pi$. The first eight terms of the solution are plotted below.

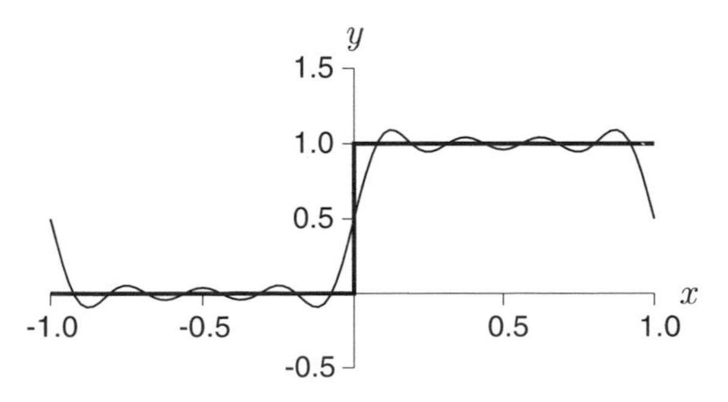

12.5.1 EVEN FOURIER SERIES

An even function $f(x)$, $x \in [-p, p]$ can be expressed as a series of cosine terms only:

$$f(x) = \frac{a_0}{2} + \sum_{n=1}^{\infty} \left(a_n \cos \frac{n\pi}{p} x \right)$$

where the coefficients a_n are given by

$$\begin{aligned} a_0 &= \frac{2}{p} \int_0^p f(x)\, dx \\ a_n &= \frac{2}{p} \int_0^p f(x) \cos \frac{n\pi}{p} x\, dx. \end{aligned}$$

EXAMPLE

The function $f(x) = |x|$, $x \in [-1, 1]$ is even and will have Fourier coefficients

$$\begin{aligned} a_0 &= 2 \int_0^1 x\, dx = 1 \\ a_n &= 2 \int_0^1 x \cos n\pi x\, dx = 2 \frac{(-1)^n - 1}{n^2 \pi^2} \end{aligned}$$

so

$$|x| = \frac{1}{2} + \sum_{n=0}^{\infty} 2 \frac{(-1)^n - 1}{n^2 \pi^2} \cos n\pi x.$$

Plotting the first four terms in the expansion gives the plot below.

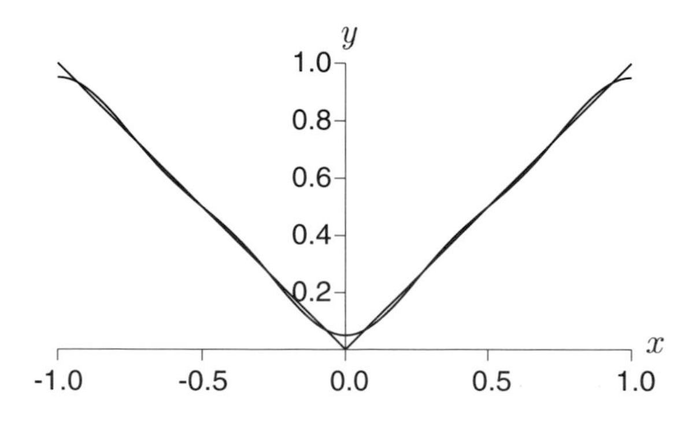

12.5.2 ODD FOURIER SERIES

An odd function $f(x)$, $x \in [-p, p]$ can be expressed as a series of sine terms only:

$$f(x) = \sum_{n=1}^{\infty} \left(b_n \sin \frac{n\pi}{p} x \right)$$

with coefficients

$$b_n = \frac{2}{p} \int_0^p f(x) \sin \frac{n\pi}{p} x \, dx.$$

EXAMPLE

The function $f(x) = x$, $x \in [-1, 1]$ is odd and will have Fourier coefficients

$$\begin{aligned} b_n &= 2 \int_0^1 x \sin n\pi x \, dx \\ &= -2 \frac{(-1)^n}{n\pi}. \end{aligned}$$

Hence

$$x = \sum_{n=0}^{\infty} -2 \frac{(-1)^n}{n\pi} \sin n\pi x.$$

Plotting the first eight terms gives the plot below.

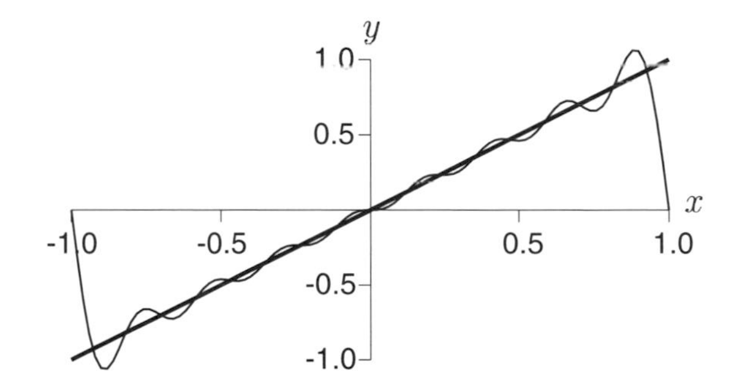

12.6 EXAMPLE QUESTIONS

(Answers are given in Chapter 14)

1. Find an approximation to the integral

$$\int_0^1 x^3\,dx$$

using an interval size of $h = 0.2$.

2. Approximate the integral

$$\int_0^2 e^x\,dx$$

using an interval size of $h = 0.5$.

3. A function $f(x)$ is defined by the set of points

$$(0,0),(1,2),(2,4),(3,7),(4,8),(5,9).$$

Hence find an approximation for the integral

$$\int_0^5 f(x)\,dx.$$

4. For the function $f(x) = \sinh x$ find an approximate value for $f'(1)$ using a discretisation $h = 0.1$ and the central differencing rule.

5. A function $f(x)$ is given by the set of points

$$(0,0),(1,1),(2,3),(3,5),(4,6),(5,7).$$

Find an approximation for $f'(x)$ at all the points $x = 0,1,2,3,4,5$.

6. Use Newton's method to find a zero of $y = x^2 - 3$ starting with $x_0 = 1$.

7. Use Newton's method to find an approximate value for $5^{1/3}$. (Hint: solve $y = x^3 - 5 = 0$ with a starting guess of $x_0 = 2$.)

8. Find the zero to two decimal places of $y = \sin x$ using Newton's method starting with $x_0 = 3$.

9. Numerically solve the differential equation

$$\frac{dy}{dx} = xy$$

with $y(0) = 1$ to find an approximation to $y(0.3)$ using a step size of $h = 0.1$.

10. Numerically solve the differential equation

$$\frac{dy}{dx} = y^2 + 1$$

with $y(0) = 0$ to find an approximation to $y(0.6)$ using a step size of $h = 0.2$.

11. Numerically solve the the equation

$$\frac{dy}{dx} = x + y^2$$

with $y(0) = 0$ to find an approximation to $y(0.4)$ using $h = 0.2$.

12. Find the Fourier series of $y = x + 1$ over the interval $x \in [-\pi, \pi]$.

13. Find the Fourier sine series of

$$f(x) = \begin{cases} -1, & -2 < x < 0 \\ 1 & 0 \le x < 2. \end{cases}$$

14. Find the Fourier cosine series for

$$f(x) = \begin{cases} 0, & -2 < x < 1 \\ 1, & -1 \le x \le 1 \\ 0, & 1 < x < 2. \end{cases}$$

15. Find the Fourier series for

$$f(x) = \begin{cases} -1, & -2 < x < -1 \\ 0, & -1 \le x \le 1 \\ 1, & 1 < x < 2. \end{cases}$$

16. A double derivative can be approximated by

$$f''(x) = \frac{f'(x+h) - f'(x-h)}{2h}.$$

If you now replace $f'(x-h)$ and $f'(x+h)$ by their respective definitions, for example

$$f'(x+h) = \frac{f(x+2h) - f(x)}{2h},$$

derive the approximation

$$f''(x) = \frac{f(x+2h) - 2f(x) + f(x-2h)}{(2h)^2}$$

or by rewriting $H = 2h$ as

$$f''(x) = \frac{f(x+H) - 2f(x) + f(x-H)}{H^2}.$$

17. For the function $f(x)$ defined by the set of points

$$(0,0),(1,2),(2,4),(3,7),(4,8),(5,9).$$

find estimates of $f'(1)$ and $f''(3)$.

CHAPTER 13
PRACTICE TESTS

Each test should be easily completed in one hour although good students will be able to do them in half an hour. A passing grade would be approximately 15 correct answers out of 20. Solutions are given in Chapter 14. The tests are only a guide and some of the more difficult areas of work may be covered at different stages in a mathematics course.

There are six tests, two for each major component of an undergraduate degree.

(i) **First Year Semester One:** Students should have a basic knowledge of algebra, functions, transcendental functions, simple differentiation and simple integration. This material is covered in Chapters 1 to 5 and it is assumed many students will know at the beginning of the semester and all students should know by the end of the semester.

(ii) **First Year Semester Two:** Students should have a more detailed knowledge of algebra, functions and transcendental functions plus differentiation (including parametric and implicit differentiation) and integration (including definite integrals, substitution and areas). A basic knowledge of vectors, matrices and asymptotics is also expected. This material is covered in Chapters 1 to 8.

(iii) **Second Year:** In addition to the previous test material the student should have a knowledge of complex numbers, integration by parts, eigenvectors, basic differential equations and multivariable calculus. Some of this material will be taught during the second year of a mathematics course depending on the university.

13.1 TEST 1: FIRST YEAR — SEMESTER ONE

1. Find the roots of the quadratic $x^2 + 1 = 5x$.
2. For what values of m is $|m + 7| > 7$?
3. Expand $\left(\frac{1}{a} + ab\right)^2$.
4. Rearrange the following equation to find y: $\frac{1}{x} + x = \frac{1}{y}$.
5. Simplify $\frac{\sqrt{xy}}{2\sqrt{y}} - \sqrt{x}$.
6. Find $\frac{dy}{dx}$ if $y = \frac{1}{2}x^3$.
7. If $f(x) = 1 - \frac{1}{x^2}$ find $f'(x)$.
8. Find $\frac{dy}{dx}$ if $y = \sin 3x^2$.
9. Evaluate $\int \frac{1}{x^3} + 1 \, dx$.
10. Find $\frac{dy}{dx}$ if $y = x \cos x$.
11. Simplify $\exp\left(\ln t^2 - 2 \ln t\right)$.
12. Simplify $\ln\left(\frac{e^x}{e^y}\right)$.
13. The equation $(x - 1)^2 + y - 3 = 0$ is what type of curve?
14. What does $f(x) = e^{-x}$ approach as $x \to -\infty$?
15. If $f(z) = 2z + 1$ and $g(x) = \frac{1}{x}$ what is $f(g(a))$?
16. Evaluate $\sum_{i=1}^{4}(i + 2)$.
17. Evaluate $\frac{6!}{4!2!}$.
18. Find A and B such that

$$\frac{1}{(x-1)(x-2)} = \frac{A}{x-1} + \frac{B}{x-2}.$$

19. Expand $(1 + x)^4$ using the binomial theorem (Pascal's triangle).
20. Divide $x^3 + 3x^2 + x + 1$ by $x + 1$.

13.2 TEST 2: FIRST YEAR — SEMESTER ONE

1. Solve for m if $m(m-4)+4=0$.

2. Factorise x^3+2x^2+x.

3. For what values of x is $|x+2|>1$?

4. Expand $(a+b+c)(a-b-c)$.

5. Rearrange $\sqrt{\dfrac{1}{x}-\dfrac{1}{y}}=3$ to find x.

6. Solve for r if $6r^2-6r-1=0$.

7. Find $\dfrac{dy}{dx}$ if $y=\dfrac{4}{x^2}$.

8. Find $f'(x)$ if $f(x)=\dfrac{1}{3}\cos 3x$.

9. Find $f'(x)$ if $f(x)=\dfrac{x}{x^2-3}$.

10. Find $f'(x)$ if $f(x)=xe^x$.

11. Evaluate $\displaystyle\int 5x^6\,dx$.

12. Evaluate $\displaystyle\int \frac{1}{2x}\,dx$.

13. If $\ln a=2$ and $\ln b=3$ what is $\ln ab^2$?

14. The equation $2x^2+y^2-3=0$ is what type of curve?

15. Evaluate the sum $\displaystyle\sum_{i=1}^{4}(2i+1)$.

16. If $f(z)=z^2$ and $g(x)=x+2$ what is $g(f(a))$?

17. Evaluate $\dfrac{5!}{3!}$.

18. Find the partial fraction form for $\dfrac{x}{(x-1)(x-3)}$.

19. Use polynomial division to divide x^3+1 by $x+1$.

20. Find the inverse function $f^{-1}(x)$ if $f(x)=\sqrt{x-2}$.

13.3 TEST 3: FIRST YEAR — SEMESTER TWO

1. Find x if $x(3x-5)=-1$.
2. For what values of x is $|x-3|<15$?
3. Expand $(3x^2-4x)^2$.
4. Rearrange $\dfrac{1}{\sqrt{y}}-\sqrt{y}=-\dfrac{1}{x}$ to find x.
5. Rearrange $\sqrt{\dfrac{1}{y}+\dfrac{1}{x}}=\dfrac{1}{9}$ to find x.
6. Find $f'(x)$ if $f(x)=3\cos(x^2)$.
7. Find $\dfrac{dy}{dx}$ if $y=x^2e^x$.
8. Find $\dfrac{dy}{dx}$ in terms of x and y if $y^2+y=\sin x^2$.
9. Find $\dfrac{dy}{dx}$ in terms of t if $y(t)=t^2$ and $x(t)=\sin t$.
10. Evaluate $\displaystyle\int \frac{1}{x-2}\,dx$.
11. Evaluate $\displaystyle\int_0^2 e^{2x}\,dx$.
12. If $f(z)=2z$ and $g(z)=1/z^2$ what is $g(f(z))$?
13. Simplify $\exp(\ln x-2\ln y)$.
14. Find the determinant
$$\begin{vmatrix} 0 & 5 & 4 \\ 3 & 0 & 2 \\ 1 & 1 & -1 \end{vmatrix}.$$
15. If $\mathbf{A}=\begin{bmatrix} 1 & 2 \\ 1 & 3 \end{bmatrix}$ what is $\mathbf{A}^{-1}$?
16. Find $\|\underset{\sim}{u}\|$ if $\underset{\sim}{u}=(-3,\sqrt{2},1)$.
17. Find $(1,2,3)\cdot(1,1,1)$.
18. Write $2\sinh x+2\cosh x$ in terms of exponentials and simplify.
19. What does $\dfrac{x+2}{x+\sqrt{x^2+3}}$ approach as $x\to\infty$.
20. How many ways are there of choosing a team of 4 people from a group of 8 people (with no order)?

13.4 TEST 4: FIRST YEAR — SEMESTER TWO

1. Solve for m if $m(m-2) = \frac{1}{2}$.
2. For what values of t is $|t-5| < 10$?
3. Expand $\left(x + \frac{1}{x}\right)^3$.
4. Make as one fraction the expression $\frac{\sqrt{x}}{2} - \frac{1}{\sqrt{x}}$.
5. Rearrange the following equation to find m: $\frac{m}{1-m} = t^2 + 3$.
6. Find $f'(x)$ if $f(x) = \frac{1}{(1-3x)^2}$.
7. Find $x'(t)$ if $x(t) = e^{-t} \sinh 2t$.
8. Find $\frac{dy}{dx}$ if $xy^2 + y = \cosh x$.
9. Find $\int \frac{1}{4x^2} + 4x^2 \, dx$.
10. The equation $x^2 + x - 2 + 3y = 0$ is what type of curve?
11. Find $\int_0^{\pi} \sin x \, dx$.
12. Find the determinant $\begin{vmatrix} 6 & -1 & 2 \\ 4 & 0 & -3 \\ 1 & 0 & -2 \end{vmatrix}$.
13. What is the angle between the vectors $(1, 2, 1)$ and $(2, -1, 0)$?
14. Find $||\underset{\sim}{u}||$ if $\underset{\sim}{u} = (2, -2, 1)$.
15. Find the partial fraction form of $\frac{1}{(x-1)^2(x-2)}$.
16. What does $f(x) = \frac{xe^x}{\sinh(x)\sqrt{x^2+2x}}$ approach as $x \to \infty$?
17. Find $\int x \sin(x^2 - 1) \, dx$ by writing $u = x^2 - 1$.
18. Find $\int_{-2}^{3} |x| \, dx$.
19. Is $f(x) = x \cosh x$ an odd or even function?
20. What is $\mathbf{A}^T\mathbf{A}$ if $\mathbf{A} = \begin{bmatrix} 1 & 2 \\ 1 & 3 \end{bmatrix}$?

13.5 TEST 5: SECOND YEAR

1. Rearrange $\sqrt{\frac{1}{y} - \frac{1}{x}} = \frac{1}{4}$ to find y.
2. Simplify $\frac{1}{\cos^2 x} - \tan^2 x$.
3. Find x if $x^2 + 5x + 6 = 0$.
4. If $\ln m = 5$ and $\ln n = 9$ find $-\ln\left(\frac{m}{n}\right)$.
5. Find $\frac{dy}{dx}$ if $y = \sin 3x^2$.
6. Find $\frac{dy}{dx}$ if $y = x^2 \sinh 2x$.
7. Find $\int \frac{1}{2x-3}\,dx$.
8. Find $\int te^{-t}\,dt$.
9. Find $\int_1^2 2x\sqrt{x^2-1}\,dx$.
10. Find $(5,2) \cdot (-7,1)$.
11. Find $(2,3,-1) \times (2,0,3)$.
12. Find the eigenvalues of $\begin{bmatrix} 6 & 5 \\ 1 & 2 \end{bmatrix}$.
13. Solve the differential equation for $y(x)$ if $\frac{dy}{dx} = -\frac{y}{x^3}$.
14. Find $y(x)$ if $\frac{d^2y}{dx^2} = -9y$.
15. Find the imaginary component of $\frac{4+2i}{3-i}$.
16. Write $z = 2 + 2i$ in polar form $z = re^{i\theta}$.
17. Find the inverse of the matrix
$$\begin{bmatrix} 1 & 0 & 1 \\ 0 & 1 & 2 \\ 0 & 1 & 1 \end{bmatrix}.$$
18. Find the first two non-zero terms of the Taylor series for $y = \sin x$ about $x = 0$.
19. What does $\frac{\cos x - 1}{x^2}$ approach as $x \to 0$?
20. If $f(x,y) = x^2 - 3y^4$ what is $\underset{\sim}{\nabla} f$?

13.6 TEST 6: SECOND YEAR

1. Factorise $2x^2 - 3x - 2$.

2. If $\ln x = 7$ and $\ln y = 2$ find $\ln\left(\frac{x^2}{y}\right)$.

3. Find $\frac{\partial f}{\partial t}$ if $f(x,t) = e^{-t}\sin \pi x$.

4. Find the imaginary component of $z = 3e^{i\pi/2}$.

5. Simplify $\frac{2x}{\sqrt{x^2 - 2x + 1}}$ as $x \to \infty$.

6. For what values of p is $|5p - 4| > 1$?

7. Simplify $\cos(n\pi)$ where $n = 0, 1, 2, \ldots$

8. Solve the differential equation $\frac{dy}{dx} = xy$.

9. Solve the differential equation $y'' + 6y' + 5y = 0$ for $y(x)$.

10. Find $\int x \cos \pi x \, dx$.

11. By letting $u = \sin x$ find $\int_{\pi/4}^{\pi/2} \frac{\cos x}{\sin^2 x}\, dx$.

12. Find $\|\underset{\sim}{\boldsymbol{u}}\|$ if $\underset{\sim}{\boldsymbol{u}} = (\sqrt{10}, 1, -1)$.

13. Find $\underset{\sim}{\boldsymbol{u}} \times \underset{\sim}{\boldsymbol{v}}$ if $\underset{\sim}{\boldsymbol{u}} = (1, 0, 1)$ and $\underset{\sim}{\boldsymbol{v}} = (0, 1, \sqrt{2})$.

14. Find the eigenvalue of $\begin{bmatrix} 2 & 1 \\ 2 & 3 \end{bmatrix}$ corresponding to the eigenvector $\underset{\sim}{\boldsymbol{v}} = (1, 2)$.

15. Find the first three terms of the Taylor series for $\exp(x^2)$ about $x = 0$.

16. If $f(x, y) = 1 - x^2 - 3y^2$ find the slope in the direction $(1, 1)$ when at the point $(0, 1)$.

17. If $f(x, y) = 1 - x^2 - 3y^2$ find the direction of maximum slope at the point $(0, 1)$.

18. Find the eigenvector of $\mathbf{A} = \begin{bmatrix} 2 & 1 \\ 1 & 2 \end{bmatrix}$ corresponding to eigenvalue $\lambda = 1$.

19. If $\underset{\sim}{\boldsymbol{v}} = (z + x, xy, 2z)$ what is $\underset{\sim}{\boldsymbol{\nabla}} \cdot \underset{\sim}{\boldsymbol{v}}$?

20. If $f = x^2 + y^3$ what is $\underset{\sim}{\boldsymbol{\nabla}}^2 f$?

CHAPTER 14

ANSWERS

Chapter 1
Algebra and Geometry

1.

(i) $-\frac{7}{30}$

(ii) $\frac{x^2 - 3x + 15}{(x+2)(x-3)}$

(iii) $\frac{x^2 + 7x - 2}{4 - x^2}$

(iv) $\frac{x^2 + 10x - 2}{(x-4)(x+2)}$

(v) $\frac{x^3 + x^2 - 3}{x(x-1)}$

(vi) $\frac{x}{x+1}$

2.

(i) $d \geq 5/2$

(ii) $d < -8$

(iii) $5 < x < 15$

(iv) $z \geq 5$ or $z \leq -11$

(v) $a > -3$ or $a < -5$

(vi) $-3 < x < 5$

3.

(i) $x^2 - 9$

(ii) $9\,x^2 - 24\,x + 16$

(iii) $x^3 + x^2y - xy^2 - y^3$

(iv) $3x^3 + 2x^2 - 27x - 18$

(v) $x^3 - 12x^2 + 48x - 64$

4.

(i) $x^4 + 8x^3 + 24x^2 + 32x + 16$

(ii) $x^8 + 8x^7 + 28x^6 + 56x^5 + 70x^4 + 56x^3 + 28x^2 + 8x + 1$

(iii) 21

5.

(i) $\frac{-3}{2}\frac{1}{x-2} + \frac{3}{2}\frac{1}{x-4}$

(ii) $\frac{1}{x-1} + \frac{3}{x+2}$

(iii) $-\frac{1}{x+3} + \frac{1}{x+2}$

(iv) $\frac{2}{x+4} + \frac{1}{x-2}$

(v) $-\frac{1}{5}\frac{1}{(x+3)^2} - \frac{1}{25}\frac{1}{x+3} + \frac{1}{25}\frac{1}{x-2}$

6.

(i) 9

(ii) $\frac{1}{3}$

(iii) 3

(iv) $\frac{3-\sqrt{3}}{3}$

7.

(i) $y = (x+5)(x+1)$

(ii) $y = (x-5)(x-1)$

(iii) $y = (x+5)(x-1)$

(iv) $y = (x-5)(x+1)$

(v) $y = (2x-1)(x+1)$

8.

(i) $x = -2, -2$

(ii) $x = -6, -1$

(iii) $x = -4, 3$

(iv) $x = -2, 1$

(v) $x = -4, 1$

(vi) $x = \frac{-1+\sqrt{13}}{2}, \frac{-1-\sqrt{13}}{2}$

9.

(i) $(x+1)+\dfrac{2}{x+2}$

(ii) $(x+1)$

(iii) x^2+3x+1

10.

(i) 720

(ii) 30

(iii) 15

(iv) 27

Chapter 2
Functions and Graphs

1. 9
2. g^3+1
3. $(x-1)^3+1$
4. $(b-1)^3+1$
5. $f(g(a))=\sin^2(a)$, $g(f(x))=\sin x^2$
6. $f(f(x))=(x^2+1)^2+1$
7. $f(g(x))=(x^2-2)^2$, $g(f(x))=(x-1)^4-1$
8. $f^{-1}(x)=\dfrac{1}{x-1}$
9. $f^{-1}(x)=\dfrac{1}{x}-1$
10. $f^{-1}(x)=\dfrac{1}{\sqrt{x-1}}$
11.

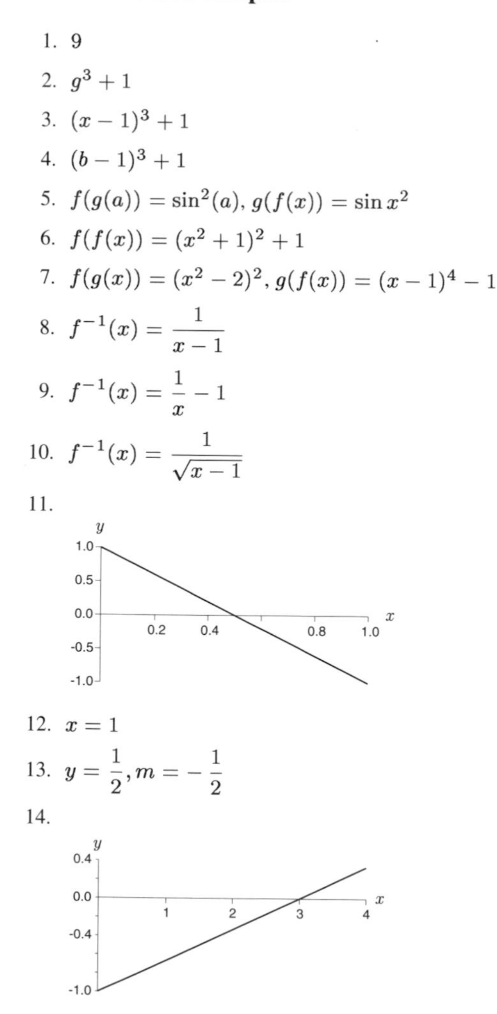

12. $x=1$
13. $y=\dfrac{1}{2}, m=-\dfrac{1}{2}$
14.

15.

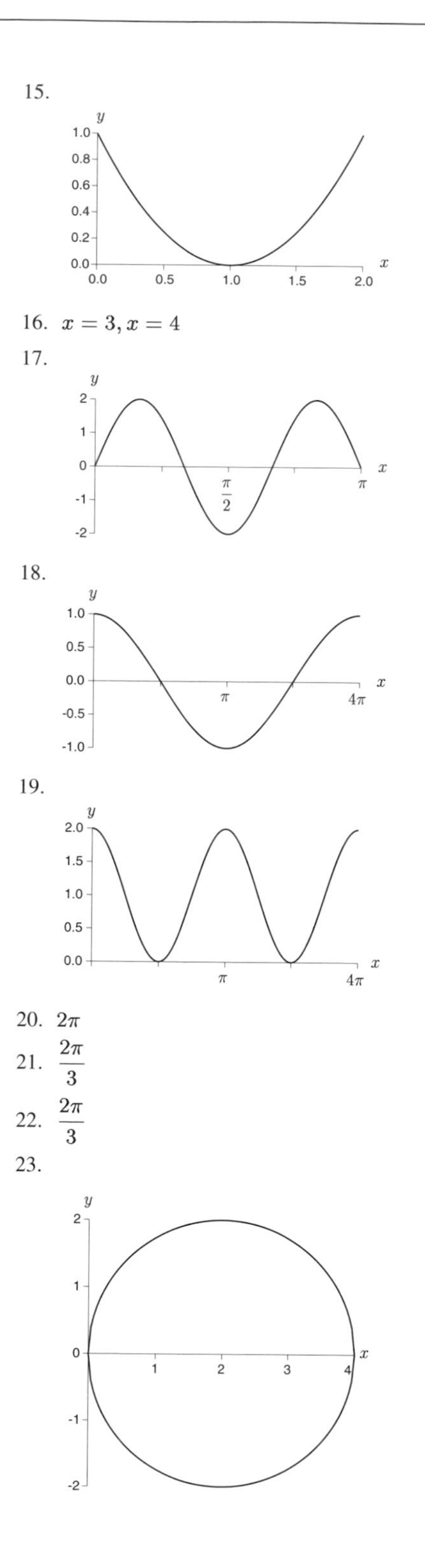

16. $x=3, x=4$
17.
18.
19.
20. 2π
21. $\dfrac{2\pi}{3}$
22. $\dfrac{2\pi}{3}$
23.

24.

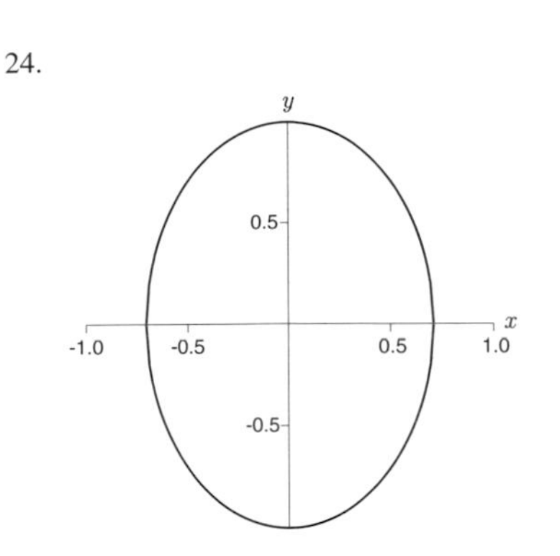

25.

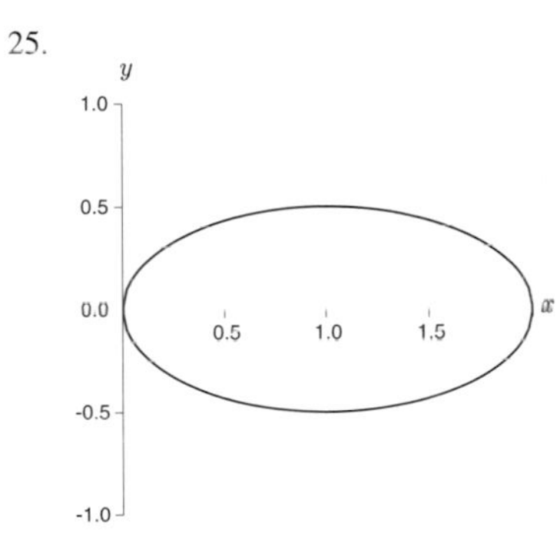

26. 2 and 0.
27. $x^2 + 4y^2 = a^2$ for some constant a
28. $(x-1)^2 + (y-2)^2 = 4$
29. $(x-a)^2 + (y-2)^2 = 9$
30. Circle, centre $(1,0)$ radius $\sqrt{2}$.
31. Ellipse, centre $(1,0)$.
32. Parabola, vertex $(1,0)$.
33. Line, gradient of $-1/2$.
34. Hyperbola, centre $(1,1)$.
35. $y = 1 + (x-2)^2$
36. $(y-1)^2 + \frac{1}{4}(x-2)^2 = 1$

Chapter 3
Transcendental Functions

1.

(i) $\frac{1}{4}\frac{y^2}{x^2}$

(ii) $\frac{1}{4}$

(iii) 2

(iv) $\frac{1}{P}$

(v) $\ln\left(\frac{x}{y}\right)$

(vi) x^2

2.

(i) $\frac{\ln 7}{\ln 5}$

(ii) $\frac{\ln 2}{\ln(1.02)}$

(iii) $\frac{\ln(5/7)}{\ln(3/2)}$

(iv) $\frac{1}{n}\frac{\ln(Q/Q_0)}{\ln(a)}$

(v) $\exp\left(\frac{3-y}{2}\right)$

(vi) $\frac{1}{4}\ln\left(\frac{3y-1}{2}\right)$

3.

(i) $\ln(st) = 2 + 3 = 5$

(ii) $\ln(st^2) = 2 + 6 = 8$

(iii) $\ln(\sqrt{st}) = (2+3)/2 = 5/2$

(iv) $\ln(s/t) = 2 - 3 = -1$

(v) $\ln(s/t^3) = 2 - 9 = -7$

4.

(i) $e^x e^y = 3 \times 5 = 15$

(ii) $e^{x+y} = e^x e^y = 15$

(iii) $e^{2x} = (e^x)^2 = 3^2 = 9$

(iv) $e^x + e^y = 3 + 5 = 8$

5.

(i) 0

(ii) $\frac{1}{\sqrt{2}}$

(iii) $\frac{\sqrt{3}}{2}$

(iv) -2

6.

(i) 1

(ii) 2

(iii) $\sin\theta$

7.

(i) $\frac{\pi}{4}, \frac{3\pi}{4}, \frac{5\pi}{4}, \frac{7\pi}{4}$

(ii) $\frac{\pi}{6}, \frac{5\pi}{6}$

8. Results given.
9. Results given.

10.

(i) $\theta = \pi/4$: $\cos\theta = 1/\sqrt{2}$, $\sin\theta = 1/\sqrt{2}$, $\tan\theta = 1$, $\sec\theta = \sqrt{2}$.

(ii) $\theta = 13\pi/6$: $\cos\theta = \sqrt{3}/2$, $\sin\theta = 1/2$, $\tan\theta = 1/\sqrt{3}$, $\sec\theta = 2/\sqrt{3}$.

(iii) $\theta = 2\pi/3$: $\cos\theta = -1/2$, $\sin\theta = \sqrt{3}/2$, $\tan\theta = -\sqrt{3}$, $\sec\theta = -2$.

(iv) $\theta = -5\pi/3$: $\cos\theta = 1/2$, $\sin\theta = \sqrt{3}/2$, $\tan\theta = \sqrt{3}$, $\sec\theta = 2$.

(v) $\theta = 5\pi/4$: $\cos\theta = -1/\sqrt{2}$, $\sin\theta = -1/\sqrt{2}$, $\tan\theta = 1$, $\sec\theta = -\sqrt{2}$.

11. $\cos\dfrac{\pi}{12} = \sqrt{(1+\sqrt{3}/2)/2}$

12. $t = 170000$

13. odd

Chapter 4 Differentiation

1.

(i) $3\cos x + 5\sin x$

(ii) $3e^x - 2x$

(iii) $\dfrac{3}{x}$

(iv) $2\cosh x - 3\sinh x$

2.

(i) $2\cos(2x)$

(ii) $(1+3x^2)\cos(x+x^3)$

(iii) $3(x+4)^2$

(iv) $5(x+\sin x)^4(1+\cos x)$

(v) $\cos(\ln x^2)\dfrac{2}{x}$

(vi) $-\exp(\cos^2 x)2\sin x\cos x$

(vii) $4x\sinh(2x^2)$

3.

(i) $e^x + xe^x$

(ii) $\dfrac{-\sin x}{x^2} - 2\dfrac{\cos x}{x^3}$

(iii) $(\cos x + \sin x)e^x$

(iv) $\dfrac{1}{x^5}(1 - 4\ln x)$

(v) $\cos^2 x - \sin^2 x$

(vi) $\left(\dfrac{1}{x} - \ln x\right)e^{-x}$

(vii) $2x\sin x + x^2\cos x$

(viii) $\sinh^2 x + \cosh^2 x$

(ix) $-e^{1/x}\left(\dfrac{1}{x^3} - \dfrac{1}{x^2}\right)$

4.

(i) $\exp(x\cos x^2)(\cos x^2 - 2x^2\sin x^2)$

(ii) $e^x(\cos(2x+1)^2 - 4(2x+1)\sin(2x+1)^2)$

(iii) $\dfrac{-x}{(2+x^2)^{3/2}}$

(iv) $\dfrac{\cos x}{(x+1)^2} - 2\dfrac{\sin x}{(x+1)^3}$

(v) $\cos(x^2 + \exp(x^3+x)) \times (2x + (3x^2+1)\exp(x^3+x))$

(vi) $2e^{x^2}\left(\dfrac{1}{x} - \dfrac{1}{x^3}\right)$

5.

(i) $\dfrac{dy}{dx} = \dfrac{\cos(x-1)}{2y} \equiv \dfrac{\cos(x-1)}{2\sqrt{\sin(x-1)}}$

(ii) $\dfrac{dy}{dx} = \dfrac{x}{2\sin(2y)\sqrt{1-x^2}} = \dfrac{1}{2}(1-x^2)^{-1/2}$

(iii) $\dfrac{dy}{dx} = e^x(x+1)y = e^x(x+1)\exp(xe^x)$

(iv) $\dfrac{dy}{dx} = \dfrac{3e^{3x}}{e^y} = \dfrac{3e^{3x}}{5+e^{3x}}$

(v) $\dfrac{dy}{dx} = \dfrac{2x}{1+3y^2}$

(vi) $\dfrac{dy}{dx} = \dfrac{\cos x}{2y+\cos y}$

(vii) $\dfrac{dy}{dx} = \dfrac{1-y}{x+1-2y}$

6.

(i) $\dfrac{dy}{dx} = -\dfrac{\sin t}{2t\cos(t^2)}$

(ii) $\dfrac{dy}{dx} = \dfrac{e^t}{2t}$

(iii) $\dfrac{dy}{dx} = \dfrac{2t}{\cos t}$

7. $\exp\left(\dfrac{a}{\sqrt{\pi t}}\right)\left(1 - \dfrac{a}{2\sqrt{\pi t}}\right)$

8.

(i) $x = 2$ minimum.

(ii) $x = 3$ minimum, $x = 1$ maximum.

(iii) $x = 0$ minimum, $x = 1$ inflection.

(iv) $x = 1$ maximum.

(v) $x = e^{-1/2}$ minimum.

(vi) $x = 0$ maximum, $x = 1$ minimum.

(vii) $x = -1$ maximum, $x = 1$ minimum.

9.

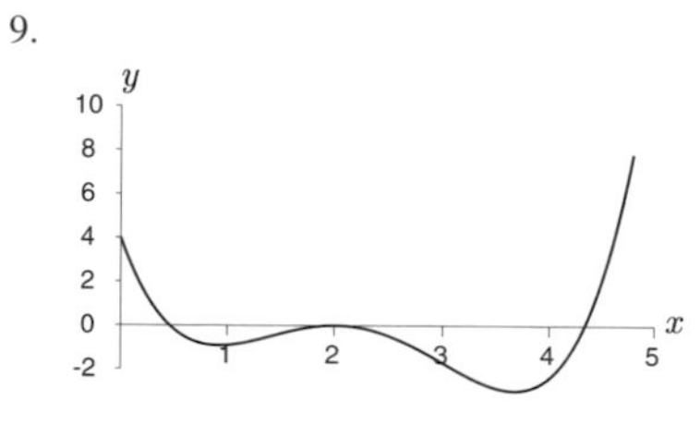

Chapter 5
Integration

1.

(i) $-\dfrac{7}{2x^2}+c$

(ii) $12\ln|x|+\dfrac{x^6}{6}+c$

(iii) $\dfrac{e^{7x}}{7}-\dfrac{e^{-14x}}{14}+c$

(iv) $\dfrac{x^{10}}{10}+\dfrac{x^{11}}{11}+\dfrac{x^{12}}{12}+\dfrac{x^{13}}{13}+c$

(v) $\dfrac{1}{2}\cosh 2x+c$

(vi) $4\sinh x-e^x+c$

2.

(i) $1-e^{-5}$

(ii) 0

(iii) $16\ln 2$

(iv) $\dfrac{2}{3}$

(v) $\dfrac{3}{\sqrt{2}}$

(vi) $7(1-e^{-3})$

3.

(i) $\dfrac{5}{2}$

(ii) $-\dfrac{1}{12}$

(iii) $-\dfrac{1}{2}$

(iv) $\dfrac{1}{2}$

4.

(i) $-\dfrac{1}{2}e^{-x^2}+c$

(ii) $\dfrac{1}{2}\ln|x^2+4x+5|+c$

(iii) $\ln|\ln y|+c$

(iv) $\dfrac{1}{21}(7x^4-1)^{3/2}+c$

(v) $-\dfrac{1}{2}\sin(-z^2)+c$

(vi) $2e^{\sqrt{s}}+c$

(vii) $\dfrac{1}{2}(1-\cos(\pi^2))$

(viii) $e^{-3}-e^{-2}$

5.

(i) -2

(ii) $-\dfrac{1}{2}\ln(3)$

(iii) $\dfrac{1}{5}e^{2x}(2\sin x-\cos x)+c$

(iv) $-(x+1)\cos x+\sin x+c$

(v) $e^x(x^2-2x+2)+c$

(vi) $x\ln x-x+c$

6.

(i) $-\ln|\cos z|+c$

(ii) $(2-x^2)\cos x+2x\sin x+c$

(iii) $9e^{x/3}(x-3)+c$

(iv) $\dfrac{2}{5}u^{5/2}+\dfrac{2}{3}u^{3/2}+c$

(v) $\operatorname{arcsinh} x+c$

Chapter 6
Matrices

1.

(i) $\mathbf{A}+\mathbf{B}=\begin{bmatrix}5&-1&0\\0&1&-9\\4&1&7\end{bmatrix}$

$\mathbf{AB}=\begin{bmatrix}7&-3&3\\-25&-15&-57\\1&-1&-1\end{bmatrix}$

$\mathbf{BA}=\begin{bmatrix}1&-3&8\\-14&-6&12\\21&3&-4\end{bmatrix}$

trace($\mathbf{A}$) = 7

(ii) $\mathbf{A}+\mathbf{B}=\begin{bmatrix}7&1\\-1&5\end{bmatrix}$

$\mathbf{AB}=\begin{bmatrix}-57&1\\8&-44\end{bmatrix}$

$\mathbf{BA}=\begin{bmatrix}-42&22\\-1&-59\end{bmatrix}$

trace($\mathbf{A}$) = 5

(iii) $\mathbf{A}+\mathbf{B}=\begin{bmatrix}11&0&0\\0&11&0\\0&0&1\end{bmatrix}$

$\mathbf{AB}=\begin{bmatrix}18&0&0\\0&24&0\\0&0&-12\end{bmatrix}$

$\mathbf{BA}=\begin{bmatrix}18&0&0\\0&24&0\\0&0&-12\end{bmatrix}$

trace($\mathbf{A}$) = 9

2.

(i) $\begin{bmatrix} 39 & 11 \\ 15 & 4 \\ 30 & 6 \end{bmatrix}$

(ii) $\begin{bmatrix} 2 & 3 & 0 & -1 \end{bmatrix}$

(iii) $\begin{bmatrix} 6 & -6 \\ -2 & 2 \end{bmatrix}$

3.

(i) $\mathbf{A}^t = \begin{bmatrix} 2 & 4 \\ -1 & 3 \\ 3 & -5 \end{bmatrix}$

$\mathbf{A}^t\mathbf{A} = \begin{bmatrix} 20 & 10 & -14 \\ 10 & 10 & -18 \\ -14 & -18 & 34 \end{bmatrix}$

$\mathbf{A}\mathbf{A}^t = \begin{bmatrix} 14 & -10 \\ -10 & 50 \end{bmatrix}$

(ii) $\mathbf{A}^t = \begin{bmatrix} 6 & 0 & 7 \\ 0 & -4 & 5 \end{bmatrix}$

$\mathbf{A}^t\mathbf{A} = \begin{bmatrix} 85 & 35 \\ 35 & 41 \end{bmatrix}$

$\mathbf{A}\mathbf{A}^t = \begin{bmatrix} 36 & 0 & 42 \\ 0 & 16 & -20 \\ 42 & -20 & 74 \end{bmatrix}$

4.

(i) $\frac{1}{5}\begin{bmatrix} 4 & -3 \\ -1 & 2 \end{bmatrix}$

(ii) $\frac{1}{2}\begin{bmatrix} 2 & -4 & 0 \\ 0 & 2 & 0 \\ 0 & -3 & 1 \end{bmatrix}$

5.

(i) -38

(ii) -36

(iii) 0

6. $x = 1, y = 1, z = 1$

7. $x = t, y = 4t, z = 16t, \qquad t \in \text{Re}$

8. (i) $a \neq 21/2$

(ii) $a = 21/2, c \neq 1$

(iii) $a = 21/2, c = 1$

9. $\lambda_1 = 2, \quad \underset{\sim}{\boldsymbol{v_1}} = (1, 0, 0)$

$\lambda_2 = 3, \quad \underset{\sim}{\boldsymbol{v_2}} = (0, 1, 0)$

$\lambda_3 = 1, \quad \underset{\sim}{\boldsymbol{v_3}} = (1, 2, -1)$

10. $\lambda_1 = 3, \quad \underset{\sim}{\boldsymbol{v_1}} = (1, 1, 0)$

$\lambda_2 = 1, \quad \underset{\sim}{\boldsymbol{v_2}} = (0, 0, 1)$

$\lambda_3 = 1, \quad \underset{\sim}{\boldsymbol{v_3}} = (1, -1, 0)$

Chapter 7
Vectors

1.

(i) $\underset{\sim}{\boldsymbol{u}} + \underset{\sim}{\boldsymbol{v}} = (-1, 0)$

$3\underset{\sim}{\boldsymbol{u}} = (-6, -3)$

$||\underset{\sim}{\boldsymbol{u}}|| = \sqrt{5}$

(ii) $(7, 7), (9, 12), 5$

(iii) $(-3, 0)\ (-6, 3), \sqrt{5}$

(iv) $(4, 5, 3), (9, 12, 6), \sqrt{29}$

(v) $(4, 1, 2, 1), (9, 3, 3, 0), \sqrt{11}$

(vi) $(3, 2, 0), (6, 9, 3), \sqrt{14}$

(vii) $(2, -2), (3, 3), \sqrt{2}$

2.

(i) $1 < \sqrt{5} + \sqrt{2}$

(ii) $7\sqrt{2} < 5 + 5$

(iii) $3 < \sqrt{5} + \sqrt{2}$

(iv) $5\sqrt{2} < \sqrt{3} + \sqrt{29}$

(v) $\sqrt{22} < \sqrt{3} + \sqrt{11}$

(vi) $\sqrt{13} < \sqrt{3} + \sqrt{14}$

(vii) $2\sqrt{2} < \sqrt{10} + \sqrt{2}$

3. $\underset{\sim}{\boldsymbol{v}} = (3, 1), \underset{\sim}{\boldsymbol{u}} = (-3, 2.5), \underset{\sim}{\boldsymbol{u}} + \underset{\sim}{\boldsymbol{v}} = (0, 3.5)$

4.

(i) $\underset{\sim}{\boldsymbol{u}} + \underset{\sim}{\boldsymbol{v}} = (2, 0, 0)$

$||\underset{\sim}{\boldsymbol{u}} + \underset{\sim}{\boldsymbol{v}}|| = 2$

(ii) $\underset{\sim}{\boldsymbol{u}} + \underset{\sim}{\boldsymbol{v}} = (-1, -2, 7)$

$||\underset{\sim}{\boldsymbol{u}} + \underset{\sim}{\boldsymbol{v}}|| = 3\sqrt{6}$

(iii) $\underset{\sim}{\boldsymbol{u}} + \underset{\sim}{\boldsymbol{v}} = (12, 3, -2)$

$||\underset{\sim}{\boldsymbol{u}} + \underset{\sim}{\boldsymbol{v}}|| = \sqrt{157}$

5.

(i) $\underset{\sim}{\boldsymbol{u}} \cdot \underset{\sim}{\boldsymbol{v}} = 6$

$\underset{\sim}{\boldsymbol{u}} \times \underset{\sim}{\boldsymbol{v}} = (-1, -2, 5)$

$\cos\theta = \sqrt{\frac{6}{11}}$

(ii) $\underset{\sim}{\boldsymbol{u}} \cdot \underset{\sim}{\boldsymbol{v}} = -17$

$\underset{\sim}{\boldsymbol{u}} \times \underset{\sim}{\boldsymbol{v}} = (3, -3, -15)$

$\cos\theta = -\frac{17}{\sqrt{14}\sqrt{38}}$

(iii) $\underset{\sim}{\boldsymbol{u}} \cdot \underset{\sim}{\boldsymbol{v}} = 11$

$\underset{\sim}{\boldsymbol{u}} \times \underset{\sim}{\boldsymbol{v}} = (-6, 4 - 10)$

$\cos\theta = \frac{11}{\sqrt{13}\sqrt{21}}$

(iv) $\underset{\sim}{\boldsymbol{u}} \cdot \underset{\sim}{\boldsymbol{v}} = 0$

$\underset{\sim}{\boldsymbol{u}} \times \underset{\sim}{\boldsymbol{v}} = \underset{\sim}{\mathbf{0}}$, undefined.

(v) $\underset{\sim}{\boldsymbol{u}} \cdot \underset{\sim}{\boldsymbol{v}} = -9$

$\underset{\sim}{\boldsymbol{u}} \times \underset{\sim}{\boldsymbol{v}} = \underset{\sim}{\mathbf{0}}$

$\cos\theta = -1$

(vi) $\underset{\sim}{\boldsymbol{u}} \cdot \underset{\sim}{\boldsymbol{v}} = 2$

$\underset{\sim}{\boldsymbol{u}} \times \underset{\sim}{\boldsymbol{v}} = (-20, 10, 0)$

$\cos\theta = \dfrac{1}{\sqrt{6}\sqrt{21}}$

6. Orthogonality easily verified.

7.

(i) independent (det=7)

(ii) dependent

(iii) dependent

(iv) independent (det=2)

(v) dependent

(vi) dependent

8. $c = -2$

9. $(1, 2, 2)$, 3

10. $(1, 2, 3) \cdot (2, -1, 0) = 0$

Chapter 8
Asymptotics and Approximations

1.

(i) 6

(ii) 1

(iii) 1

(iv) 0

(v) $\dfrac{1}{2}$

(vi) 0

2.

(i) $-\dfrac{1}{6}$

(ii) 0

(iii) $\dfrac{1}{3}$

(iv) $-\dfrac{1}{3}$

3.

(i) $\cos x \approx 1 - \dfrac{x^2}{2}$

(ii) $\dfrac{\sin x}{x} \approx 1 - \dfrac{x^2}{6}$

(iii) $\dfrac{x^2+1}{1+1/x} \approx x - x^2$

(iv) $xe^x \approx x + x^2$

(v) $\dfrac{\cos x - 1}{x} \approx -\dfrac{x}{2} + \dfrac{1}{24}x^3$

(vi) $\sinh x \approx x + \dfrac{1}{6}x^3$

(vii) $\dfrac{\cosh x - 1}{x^2} \approx \dfrac{1}{2} + \dfrac{1}{24}x^2$

(viii) $\dfrac{1}{1-x} \approx 1 + x$

(ix) $\ln(1+x) \approx x - \dfrac{1}{2}x^2$

4. As $x \to \infty$:

(i) $\dfrac{x^2+x}{3x^2+2x+1} \approx \dfrac{1}{3}$

(ii) $\sqrt{x^3+2x+1} \approx x^{3/2}$

(iii) $\dfrac{x^3+2x}{3x^2+\sqrt{x^4+1}} \approx \dfrac{x}{4}$

(iv) $\dfrac{x^2+1}{1+1/x} \approx x^2$

(v) $\dfrac{\sinh(x)}{\cosh(x)} \approx 1$

(vi) $\dfrac{\sinh(3x)}{1+e^{4x}} \approx \dfrac{e^{-x}}{2}$

(vii) $\dfrac{xe^{-x}}{\sinh x} \approx 2xe^{-2x}$

(viii) $\dfrac{x+e^{-x}}{1+xe^{-x}} \approx x$

(ix) $\dfrac{e^x}{\sinh x} \approx 2$

5.

(i) $e^x \sin x \approx x + x^2 + \frac{1}{3}x^3$

(ii) $e^{2x} \approx 1 + 2x + 2x^2$

(iii) $\sin x^2 \approx x^2 - \frac{1}{6}x^6 + \frac{1}{120}x^{10}$

(iv) $e^{x^3} \approx 1 + x^3 + \frac{1}{2}x^6$

6.

(i) $\sin x \approx 1 - \frac{1}{2}\left(x - \frac{\pi}{2}\right)^2$

(ii) $\cos x \approx -\left(x - \frac{\pi}{2}\right) + \frac{1}{6}\left(x - \frac{\pi}{2}\right)^3$

7. $\displaystyle\int_0^\epsilon e^{-x^2}\,dx \approx \epsilon - \frac{1}{3}\epsilon^3 + \frac{1}{10}\epsilon^5$

Chapter 9
Complex Numbers

1.

(i) $6-25i$

(ii) $12+6i$

(iii) $82-33i$

(iv) $8-i$

(v) $13+84i$

(vi) $-2-2i$

(vii) $34(1+i)$

(viii) 2

(ix) 13

(x) $6-7i$

(xi) $\frac{26}{53}+\frac{38}{53}i$

(xii) $-i$

(xiii) $\frac{1}{2}(1-i)$

(xiv) $\frac{37}{145}-\frac{9}{145}i$

2.

(i) $x=4, y=2$

(ii) $x=3$

3.

(i) $\operatorname{cis}\pi$

(ii) $\operatorname{cis}3\pi/2$

(iii) $3\operatorname{cis}\frac{\pi}{2}$

(iv) $\sqrt{2}\operatorname{cis}\frac{3\pi}{4}$

(v) $\sqrt{2}\operatorname{cis}\frac{\pi}{4}$

(vi) $\sqrt{2}\operatorname{cis}\frac{7\pi}{4}$

(vii) $2\sqrt{2}\operatorname{cis}\frac{5\pi}{4}$

(viii) $5\operatorname{cis}\frac{11\pi}{6}$

(ix) $2\operatorname{cis}\frac{\pi}{3}$

(x) $2\sqrt{3}\operatorname{cis}2\pi/3$

(xi) $5\sqrt{2}\operatorname{cis}\frac{\pi}{4}$

(xii) $2\operatorname{cis}\frac{\pi}{6}$

(xiii) $2\sqrt{2}\operatorname{cis}\frac{7\pi}{6}$

4.

(i) $2i$

(ii) $\frac{3}{\sqrt{2}}(1+i)$

(iii) $\frac{1}{2}(\sqrt{3}+i)$

(iv) $\frac{1}{2}(\sqrt{3}-i)$

(v) $\frac{1}{2}(1+\sqrt{3}I)$

5.

(i) $z_1z_2=e^{3\pi/4}=\frac{1}{\sqrt{2}}(-1+i)$,
$z_1^2=e^{i\pi}=-1$

(ii) $z_1z_2=e^{i\pi/2}=i$,
$z_1^2=e^{i\pi/3}=\frac{1}{2}(1+\sqrt{3}i)$

(iii) $z_1z_2=6e^{i\pi}=-6$,
$z_1^2=4e^{i\pi/2}=4i$

(iv) $z_1z_2=3e^{-2i\pi/3}=\frac{3}{2}(-1-\sqrt{3}i)$,
$z_1^2=e^{i\pi/3}=\frac{1}{2}(1+\sqrt{3}i)$

6.

(i) $e^{i\pi/4}=\frac{1}{\sqrt{2}}(1+i)$
$e^{i5\pi/4}=-\frac{1}{\sqrt{2}}(1+i)$

(ii) $\pm1, \pm i$

(iii) $2e^{i\pi/4}, 2e^{i11\pi/12}, 2e^{-i5\pi/12}$

(iv) $\frac{1}{\sqrt{2}}(1-i), \frac{1}{\sqrt{2}}(-1+i)$

(v) $\operatorname{cis}(i(\pi/6+k\pi/3)), k=0,\ldots,5$ or
$z=\pm i, \frac{1}{2}(\sqrt{3}\pm i), \frac{1}{2}(-\sqrt{3}\pm i)$

Chapter 10
Differential Equations

1.

(i) $\frac{1}{3}(x+1)^3+c$

(ii) $-\frac{1}{x}+c$

(iii) $y=x-\ln|x+1|+c$

(iv) $y=\frac{1}{5}\ln x+c$

(v) $y=3x\cos x-3\sin x+c$

2.

(i) $x=\frac{1}{4}y^4+c$

(ii) $\frac{1}{x}-\frac{1}{2y^2}=c$

(iii) $y=cx-1$

(iv) $-3e^{-2y}=2e^{3x}+c$

(v) $\ln|y+1| + \dfrac{1}{y+1} = \dfrac{1}{2}\ln\left|\dfrac{x+1}{x-1}\right| + c$

(vi) $-\cos x = \ln|y| + y^2 + c$

(vii) $-x\cos x + \sin x = ye^y - e^y + c$

(viii) $\dfrac{1}{12}(4x+5)^3 - \dfrac{1}{6}(2y+3)^3 = c$

3.

(i) $y = \dfrac{1}{10} + ce^{-5x}$

(ii) $y = \dfrac{3}{2} + \dfrac{c}{x^2}$

(iii) $y = \dfrac{1}{2} + ce^{-x^2}$

(iv) $y = \dfrac{1}{3} + ce^{-x^3}$

(v) $y = \sin x + c\cos x$

(vi) $y = \dfrac{c}{x^3 - 1}$

4.

(i) $Q = Ce^{kt} + 70$

(ii) $y = \dfrac{c}{e^x + 1}$

(iii) $r = \dfrac{\theta - \cos\theta + c}{\sec\theta + \tan\theta}$

(iv) $P(t) = \dfrac{1}{1 + ce^{-t}}$

(v) $y = \pm\sqrt{c + x + x^2}$

(vi) $i = \dfrac{E}{R} + ce^{-Rt/L}$

5.

(i) $y(x) = c_1 e^{4x} + c_2 e^{-4x}$

(ii) $y(x) = e^{-2x}(c_1 e^{-\sqrt{5}x} + c_2 e^{\sqrt{5}x})$

(iii) $y(x) = c_1 e^{2x/3} + c_2 e^{-x/4}$

(iv) $y(x) = c_1 \cos 3x + c_2 \sin 3x$

(v) $y(x) = c_1 e^{-4x} + c_2 x e^{-4x}$

(vi) $y(x) = c_1 e^{-x/2} + c_2 e^{x/4}$

(vii) $y(x) = c_1 + c_2 e^{-5x/2}$

(viii) $y(x) = e^{-x/3} c_1 \cos \dfrac{\sqrt{2}}{3} x$
$+ e^{-x/3} c_2 \sin \dfrac{\sqrt{2}}{3} x$

(ix) $y(x) = c_1 e^{5x} + c_2 x e^{5x}$

6.

(i) $y(x) = c_1 e^{5x} + c_2 e^{-2x} + \dfrac{3}{10}$

(ii) $y(x) = 6 - 4x + x^2 + c_1 e^{-x} + c_2 x e^{-x}$

(iii) $y(x) = c_1 \cos x + c_2 \sin x - \dfrac{1}{8}\cos 3x$

(iv) $y(x) = \dfrac{1}{5} e^{2x} + c_1 \cos x + c_2 \sin x$

(v) $y(x) = -\dfrac{24}{25}\cos 2x - \dfrac{18}{25}\sin(2x) + c_1 e^{-x} + c_2 x e^{-x}$

(vi) $y(x) = c_1 e^x + c_2 e^{-x} + \dfrac{1}{2} x e^x - \dfrac{1}{4} e^x - x^2 - 2$

Chapter 11
Multivariable Calculus

1.

(i) $\dfrac{\partial f}{\partial x} = 2y$
$\dfrac{\partial f}{\partial y} = 2y + 2x$
$\dfrac{\partial^2 f}{\partial x \partial y} = 2$

(ii) $\dfrac{\partial f}{\partial x} = y\cos xy$
$\dfrac{\partial f}{\partial y} = x\cos xy,$
$\dfrac{\partial^2 f}{\partial x \partial y} = \cos xy - xy\sin xy$

(iii) $\dfrac{\partial f}{\partial x} = 2x - \sin(x+y)$
$\dfrac{\partial f}{\partial y} = -\sin(x+y)$
$\dfrac{\partial^2 f}{\partial x \partial y} = -\cos(x+y)$

(iv) $\dfrac{\partial f}{\partial x} = e^{x-y}$
$\dfrac{\partial f}{\partial y} = -e^{x-y}$
$\dfrac{\partial^2 f}{\partial x \partial y} = -e^{x-y}$

(v) $\dfrac{\partial f}{\partial x} = 3yx^2$
$\dfrac{\partial f}{\partial y} = x^3$
$\dfrac{\partial^2 f}{\partial x \partial y} = 3x^2$

2.

(i) $(2x + 2z, -2y, 2x)$

(ii) $\sec^2(xyz)(yz, xz, xy)$

(iii) $\dfrac{1}{2\sqrt{y}} + 11x$

(iv) $(x(z-y), y(x-z), z(y-x))$

(v) $1 + 2 + 3 = 6$

(vi) $2x + 4y + 6z$

(vii) $2 + 2 + 2 = 6$

(viii) $2(yz)^2 + 2(xz)^2 + 2(xy)^2$

(ix) $(0, 0, 2x - 2y)$

(x) $(2y, 2z, 2x)$

3.

(i) $(4, 3)$

(ii) 5

(iii) $\pm(3, -4)$

(iv) $\frac{7}{\sqrt{2}}$

4.

(i) $(-1, -1)$

(ii) $\sqrt{2}$

(iii) $\pm(1, -1)$

(iv) $-\sqrt{2}$

5. $4t^3 - 3t^2$

6.

(i) 54

(ii) $\frac{64}{3}$

(iii) 416

7. $\int_{x=0}^{1} \int_{y=x^2}^{1} x^2 + y^2 \, dy dx = \frac{44}{105}$

8. $\int_{y=0}^{1} \int_{x=0}^{\sqrt{y}} x^2 + y^2 \, dx dy = \frac{44}{105}$

9. $\int_{x=0}^{1} \int_{y=x}^{\sqrt{x}} x + y \, dy dx = \frac{3}{20}$

10. $\int_{y=0}^{1} \int_{x=y^2}^{y} x + y \, dy dx = \frac{3}{20}$

11. Proof by expansion.

12. $\frac{\partial^2 c}{\partial x^2} = \frac{\partial c}{\partial t} = \frac{\exp\left(-\frac{x^2}{4t}\right)(x^2 - 2t)}{4t^{5/2}}$

Chapter 12
Numerical Skills

1. $\int_0^1 x^3 \approx 0.26$

2. $\int_0^2 e^x \, dx \approx 6.52.$

3. $\int_0^5 f(x) \, dx \approx 25.5$

4. $f'(1) \approx 1.55$

5. $f'(0) \approx 1, f'(1) \approx 1.5, f'(2) \approx 2,$
$f'(3) \approx 1.5, f'(4) \approx 1, f'(5) \approx 1$

6. $x_0 = 1, x_1 = 2, x_2 = 1.75, x_3 = 1.732$

7. $x_0 = 2, x_1 = 1.75, x_2 = 1.711, x_3 = 1.70997$

8. $x_0 = 3, x_1 = 3.1425, x_2 = 3.14159$

9. $y(0.3) \approx 1.03$

10. $y(0.6) \approx 0.6$

11. $y(0.4) \approx 1.5$

12. $a_0 = 2, a_n = 0, b_n = \frac{-2(-1)^n}{n}$

13. $b_n = 2\frac{1-(-1)^n}{n\pi}$

14. $a_0 = 1, a_n = 2\frac{\sin(n\pi/2)}{n\pi}$

15. $b_n = \frac{1}{n\pi}\left(\cos\frac{n\pi}{2} - (-1)^n\right), a_n = 0$

16. Easily proved.

17. $f'(1) = 2, f''(3) = -2$

Chapter 13
Practice Tests

Test 1: Section 13.1

1. $x = \frac{5 \pm \sqrt{21}}{2}$

2. $m > 0$ or $m < -14$

3. $a^2b^2 + 2b + \frac{1}{a^2}$

4. $\frac{x}{1+x^2}$

5. $-\frac{\sqrt{x}}{2}$

6. $\frac{3x^2}{2}$

7. $\frac{2}{x^3}$

8. $6x \cos 3x^2$

9. $x - \frac{1}{2x^2}$

10. $\cos x - x \sin x$

11. 1

12. $x - y$

13. Parabola.

14. ∞

15. $\frac{2}{a} + 1$

16. 18

17. 15

18. $A = -1, B = 1$

19. $x^4 + 4x^3 + 6x^2 + 4x + 1$

20. $x^2 + 2x - 1 + \frac{2}{x+1}$

Test 2: Section 13.2

1. $m = 2$
2. $x(x+1)^2$
3. $x < -3$ or $x > -1$
4. $a^2 - b^2 - 2bc - c^2$
5. $\dfrac{y}{1+9y}$
6. $\dfrac{1}{2} \pm \dfrac{\sqrt{15}}{6}$
7. $-\dfrac{8}{x^3}$
8. $-\sin 3x$
9. $-\dfrac{x^2+3}{(x^2-3)^2}$
10. $e^x + xe^x$
11. $\dfrac{5x^7}{7}$
12. $\dfrac{1}{2}\ln x$
13. 8
14. Ellipse
15. 24
16. $a^2 + 2$
17. 20
18. $\dfrac{1}{2}\dfrac{1}{x-1} + \dfrac{3}{2}\dfrac{1}{x-3}$
19. $x^2 - x + 1$
20. $f^{-1}(x) - 2 + x^2$

Test 3: Section 13.3

1. $\dfrac{1}{6}(5 \pm \sqrt{13})$
2. $-12 < x < 18$
3. $9x^4 - 24x^3 + 16x^2$
4. $-\dfrac{\sqrt{y}}{y-1}$
5. $\dfrac{81y}{y-81}$
6. $-6x \sin x^2$
7. $2xe^x + x^2e^x$
8. $\dfrac{2x\cos x^2}{1+2y}$
9. $\dfrac{2t}{\cos t}$
10. $\ln(x-2)$
11. $\dfrac{1}{2}(e^4 - 1)$
12. $\dfrac{1}{4z^2}$
13. $\dfrac{x}{y^2}$
14. 37
15. $\begin{bmatrix} 3 & -2 \\ -1 & 1 \end{bmatrix}$
16. $2\sqrt{3}$
17. 6
18. $2e^x$
19. $\dfrac{1}{2}$
20. $\dfrac{8!}{4!4!} = 70$

Test 4: Section 13.4

1. $1 \pm \dfrac{\sqrt{6}}{2}$
2. $-5 < t < 15$
3. $x^3 + 3x + \dfrac{3}{x} + \dfrac{1}{x^3}$
4. $\dfrac{x-2}{2\sqrt{x}}$
5. $\dfrac{t^2+3}{4+t^2}$
6. $\dfrac{6}{(1-3x)^3}$
7. $e^{-t}(2\cosh 2t - \sinh 2t)$
8. $\dfrac{\sinh x - y^2}{1 + 3yx}$
9. $-\dfrac{1}{4x} + \dfrac{4}{3}x^3$
10. Parabola
11. 2
12. -5
13. $\dfrac{\pi}{2}$
14. 3
15. $\dfrac{1}{x-2} - \dfrac{1}{x-1} - \dfrac{1}{(x-1)^2}$
16. 2
17. $-\dfrac{1}{2}\cos(x^2 - 1)$
18. $\dfrac{13}{2}$
19. odd
20. $\begin{bmatrix} 2 & 5 \\ 5 & 13 \end{bmatrix}$

Test 5: Section 13.5

1. $\dfrac{16x}{16+x}$
2. 1
3. $x = -2, 3$
4. 4
5. $6x \cos 3x^2$
6. $2x \sinh 2x + 2x^2 \cosh 2x$
7. $\dfrac{1}{2} \ln(2x-3)$
8. $-e^{-t}(t+1)$
9. $2\sqrt{3}$
10. -33
11. $(9, -8, -6)$
12. $\lambda = 7, \lambda = 1$
13. $y = c_1 \exp\left(\dfrac{1}{2x^2}\right)$
14. $y = c_1 \sin 3x + c_2 \cos 3x$
15. i
16. $2\sqrt{2} \operatorname{cis} \dfrac{\pi}{4}$
17. $\begin{bmatrix} 1 & -1 & 1 \\ 0 & -1 & 2 \\ 0 & 1 & -1 \end{bmatrix}$
18. $\sin x \approx x - \dfrac{x^3}{6}$
19. $-\dfrac{1}{2}$
20. $(2x, -12y^3)$

Test 6: Section 13.6

1. $(2x+1)(x-2)$
2. 12
3. $-e^{-t} \sin \pi x$
4. $3i$
5. 2
6. $p > 1$ or $p < \dfrac{3}{5}$
7. $(-1)^n$
8. $y = c_1 \exp\left(\dfrac{x^2}{2}\right)$
9. $y = c_1 e^{-x} + c_2 e^{-5x}$
10. $\dfrac{1}{\pi^2}(\cos \pi x + \pi x \sin \pi x)$
11. $\sqrt{2} - 1$
12. $2\sqrt{3}$
13. $(-1, -\sqrt{2}, 1)$
14. $\lambda = 4$
15. $e^{x^2} \approx 1 + x^2 + \dfrac{x^4}{2}$
16. $-3\sqrt{2}$
17. $(0, -6)$
18. $(1, -1)$
19. $3 + x$
20. $2 + 6y$

CHAPTER 15

OTHER ESSENTIAL SKILLS

Use these pages to write any other essential mathematics skills your lecturers think are appropriate.

Index